SpringerBriefs in Applied Sciences and Technology

Computational Intelligence

Series editor

Janusz Kacprzyk, Warsaw, Poland

More information about this series at http://www.springer.com/series/10618

Inés Couso · Didier Dubois
Luciano Sánchez

Random Sets and Random Fuzzy Sets as Ill-Perceived Random Variables

An Introduction for Ph.D. Students and Practitioners

Springer

Inés Couso
Departamento de Estadística e I. O. y D. M.
Universidad de Oviedo
Oviedo
Spain

Luciano Sánchez
Departamento de Informática
Universidad de Oviedo
Oviedo
Spain

Didier Dubois
IRIT, CNRS
Université Paul Sabatier
Toulouse
France

ISSN 2191-530X
ISBN 978-3-319-08610-1
DOI 10.1007/978-3-319-08611-8

ISSN 2191-5318 (electronic)
ISBN 978-3-319-08611-8 (eBook)

Library of Congress Control Number: 2014942636

Springer Cham Heidelberg New York Dordrecht London

Printed on acid-free paper

Springer is part of Springer Science+Business Media (www.springer.com)

Preface

Several books written by prominent authors during the last 40 years have covered different aspects of random sets rather independently of one another. In most cases, random sets are considered as precise representations of random complex entities whose outcomes were naturally described by sets of elements. Independently, the theory of belief functions uses a similar formal setting to describe an extension of subjective probabilities that accommodates incomplete information. Besides, the recent literature contains many papers about fuzzy random variables, that often adopt the first point of view. This book is intended to focus on the use of random sets and fuzzy random variables as natural representations of ill-observed variables or partially known conditional probabilities.

The content of this book is based on several joint chapters written by the authors, enlarged and completed with examples and exercises. It deals with the use of random sets and fuzzy random variables from an epistemic point of view. This is closely connected to the possibilistic interpretation of fuzzy sets, suggested by L. A. Zadeh in 1978. Within this context, the relation between possibility measures and families of nested confidence intervals, and their relation with cuts of fuzzy sets was independently studied by Inés Couso and Luciano Sánchez, and Didier Dubois and colleagues during the last decade. Later on, a joint study about the combination of possibility and probability measures, carried out in the context of the Ph.D. thesis of Cédric Baudrit, could shed light on various related issues, especially connecting Dempster upper and lower probabilities to fuzzy random variables, and questioning the relevance of classical approaches proposed to evaluate statistical parameters in this setting.

The terms "ontic" and "epistemic" fuzzy sets were suggested by D. Dubois and H. Prade in order to distinguish between complex set-valued entities and the representation of partial knowledge about point-valued entities. This distinction naturally leads to three different interpretations of fuzzy random variables, one of them related to the ontic interpretation of fuzzy sets, and the remaining two related to the epistemic one. An initial joint work in the context of the Ph.D. thesis by Cédric Baudrit served to deepen the study of this categorization. Long discussions between the authors about possible meanings of sets and fuzzy sets, led them to propose three different interpretations of fuzzy random variables, that are present in this book.

The authors are indebted to other colleagues for numerous discussions. Serafín Moral and Gert de Cooman were kind enough to review the work of Inés Couso in early stages of her research activity, and to suggest interesting ideas. Some discussions with them about objective and subjective views of probabilities and previsions were very enriching. Debates with María Ángeles Gil, Ana Colubi, and Reinhard Viertl about the nature of fuzzy set-valued data were also fruitful. Part of those debates took place at the first Workshop on Harnessing the Information Contained in Low Quality Data that took place in Mieres (Spain) in 2012, and are reported in a Special Issue of the *International Journal of Approximate Reasoning*, edited by Inés Couso and Luciano Sánchez. During the final stages of her Ph.D. studies, Inés Couso had the opportunity to speak with Peter Walley about some of the aspects about information preserved by nonconvex sets of probabilities induced by random sets and the loss of relevant information about certain parameters involved in the convexification of such a family of probabilities. They also discussed (along with Serafín Moral) about different aspects of the notions of conditioning and independence in the context of imprecise probabilities.

This work has been partially supported by the Spanish project TIN2011-24302.

Contents

Chapter 1
Introduction

Random sets originate in works published in the mid-sixties by well-known economists, Aumann [1] and Debreu [3] on the integration of set-valued functions. They have been given a full-fledged mathematical development by Kendall [10] and Matheron [16]. Random sets seem to have been originally used to handle uncertainty in spatial information, namely to tackle uncertainty in the definitions of geographical areas, in mathematical morphology, and in connection to geostatistics (to which Matheron is a major pioneering contributor, as seen by his work on kriging). Under this view, a precise realisation of a random set process is a precisely located set or region in an area of interest. This approach, especially applied to continuous spaces, raises subtle mathematical issues concerning the correct topology for handling set-valued realisations, that perhaps hide the intrinsic simplicity and intuitions behind random sets (e.g., casting them in a finite setting). The reason is that continuous random sets, like in geostatistics, were perhaps more easily found in applications than finite ones at that time. In any case this peculiarity has confined random sets to very specialised areas of mathematics.

Yet, as argued in this book, random sets, including and especially finite ones, can be useful in other areas, and especially information processing and knowledge representation. More precisely, the treatment of incomplete or imprecise statistical data can benefit from random sets. Indeed, sets can represent incomplete information about otherwise precise variables, that is, a random set can be a natural model for an ill-known random variable. A random set is then a natural representation of a set of possible random variables, one of which is the right one, what we can call the *epistemic* understanding of random sets. This kind of view is at work in the pioneering works of Dempster [4] who studied upper and lower probabilities induced by a multivalued mapping from a probability space to the range of an attribute of interest. That multivalued mapping is formally the same as a random set, but now, the set has a different meaning from the one in geostatistics. In the latter area, there is no such things as upper and lower probabilities. However, if the random variable is ill-known, the probabilities of events become ill-known as well. Dempster seems to have been interested in catching up with older debates on the meaning of probabilities and of the likelihood functions, in which the statistician Ronald Fisher was involved in the

I. Couso et al., *Random Sets and Random Fuzzy Sets as Ill-Perceived Random Variables*, SpringerBriefs in Computational Intelligence, DOI: 10.1007/978-3-319-08611-8_1

first half of the twentieth century, and that were left unsolved.[1] In Dempster's view, if observable quantities can be related to the ill-known parameter characterising a probabilistic model via a function with a known probability distribution, observations generate a random set of possible parameter values for the model. On this ground, Shafer [19] developed his theory of evidence, breaking loose from the statistical setting of Dempster upper and lower probabilities, and considering them as a form of non-additive subjective probabilities that go back to some ill-studied part of Bernoulli's works. The mathematical building block of evidence theory is a probability distribution on the powerset of a finite set. The ensuing popularity of Shafer theory in artificial intelligence showed that there could be plenty of applications for finite random sets. Strangely enough there seems to have been very few cross-references from the Dempster–Shafer literature to the random set literature and conversely, if we except Nguyen's early 1978 paper pointing out the formal similarities between random sets and belief functions [17]. In this publication, the basic mathematics of random sets are discussed in detail in an elementary framework, with due credit to both traditions.

The emergence of fuzzy sets after Lotfi Zadeh's pioneering paper in 1965 [26][2] led to their hybridisation with random sets, first by the French mathematician Robert Féron in 1976, with his paper [9] in the Comptes Rendus de l'Académie des Sciences, written in the French language, and whose title strictly speaking means "random fuzzy sets". This paper and other more extensive ones (in a little known French mathematical economics outlet) appeared just after the introduction of random sets in geostatistics by another French pioneer, Matheron, and they are in the same spirit.

In 1978 and 1979, Kwakernaak [14, 15] proposed the notion of fuzzy random variable completely independently from the random set tradition, with a clear intention to represent an ill-known random variable (he explicitly mentions the fact that behind a fuzzy random variable there is an original one that is standard, but ill-perceived). Indeed, Kwakernaak does not refer to random sets at all, nor to Dempster's works. While Féron was fuzzifying random sets, Kwakernaak did randomise fuzzy intervals understood as incomplete information.

The epistemic stance of Kwakernaak approach is testified by the fact that his second paper [15] cites the seminal paper by Zadeh on possibility theory [27]. Zadeh's paper prompted Dubois and Prade to relate fuzzy sets, viewed as possibility distributions representing incomplete information, and Shafer belief functions. In their 1980 book [6], they point out that Zadeh's *possibility measures* are consonant plausibility functions, and introduce the name *necessity measures* for the conjugate functions of possibility measures, which are special case of belief functions. This book, along with some subsequent publications by Yager [23, 24] contributed to highlight possibility theory as an elementary basic building block of uncertainty representations. Consonant versions of Shafer's belief and plausibility functions

---

[1] In his paper, Dempster refers to the early works of Smith [21] on imprecise probabilities, but no reference to Aumann and Debreu is made (in fact these pioneers work more or less at the same time in different areas).

[2] Interestingly, the very same year as Auman and Debreu's papers.

actually date back to pioneering works from the late 1940s on by the economist Shackle [20], who should be considered as the forerunner of possibility theory. Possibility theory received an extensive treatment in [8], and has since then been acknowledged as one of the three major uncertainty theories along with Shafer's evidence theory and the one of imprecise probabilities [22].

The objectivist tradition was taken over in the 1980s, by Puri and Ralescu [18] who provide a rigorous mathematical foundation for fuzzy random sets, pursuing the line opened by Féron. They indeed seem to cast their work in the Auman-Debreu-Matheron tradition of random sets (even if citing many above-mentioned forerunners, with the notable exception of Dempster and Shafer and possibility theory). They started to extend standard probabilistic notions to this setting (expectation, limit theorems, normality...), while, much later on, Körner [11] construes the variance of a fuzzy random variable as a precise number. On the contrary, the book by Kruse and Meyer [13], whose works parallel the ones of Puri and Ralescu, catches up with the tradition, initiated by Kwakernaak, of viewing a fuzzy random variable as a tool for modeling incomplete fuzzy (they call it vague) information. As a consequence they consider the variance of a fuzzy random variable as a fuzzy set modeling what is known about the variance of the original random variable. Actually the difference between the two traditions cannot be observed by studying the expected value, which in the two traditions, is an interval or a fuzzy interval in the sense of Aumann integral. But the way the variance is defined[3] is a good indication of whether a random (fuzzy) set models incomplete information or not. Figure 1.1 provides an overview of the history of ideas and formal concepts outlined in this introduction.

Strangely enough, in the 1990s and later, many mathematical contributions to the theory of fuzzy random variables (see for instance [2]) followed the Puri and Ralescu tradition, not so much the Kwakernaak–Kruse one, even if, in these subsequent works, there was a deliberate intention of relating the mathematical developments and the ensuing statistical tools to ill-perceived outcomes to random experiments taking the form of linguistic variables in the sense of Zadeh, hence fuzzy numbers.[4] This is perhaps due to the lack of awareness of the existence of two distinct epistemological traditions, beyond the mathematical differences pertaining to the kind of metric spaces used, etc.

The aim of this book is to contribute to highlight the distinction between ontic and epistemic random sets, and presents the basics of the epistemic approach in connexion to imprecise probability theory, thus bridging the gaps between the works of the Kwakernaak–Kruse tradition, and Dempster's pioneering works in upper and lower probabilities. Our claim is that while the Puri–Ralescu "objectivist" tradition seems to be fit to the modeling of the variability of entities naturally taking the form of fuzzy sets, the question of handling epistemic uncertainty in statistical processes

---

[3] Compare the papers by Körner [11] and Kruse [12].

[4] In parallel, some works proposed a direct fuzzification of Shafer's theory of evidence (originating quite early in a paper by Zadeh [28]) such as the papers of Dubois and Prade [7] (also relying on Dempster's construction), Yen [25], or more recently Denœux [5]; these authors make no reference to fuzzy random variables.

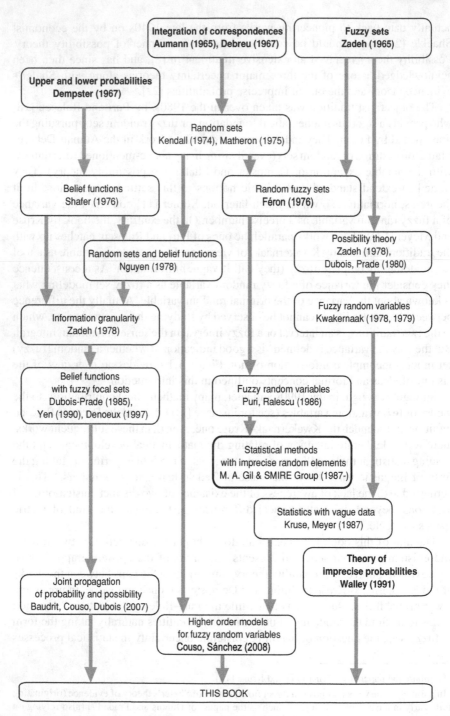

Fig. 1.1 History of random and fuzzy set representations of uncertainty

is better addressed following the Kwakernaak–Kruse tradition, which can as well be viewed as blending possibility and probability theories.

References

1. J. Aumann, Integral of set valued functions. J. Math. Anal. Appl. **12**, 1–12 (1965)
2. A. Colubi, R. Coppi, P. D'Urso, M.A. Gil. Statistics with fuzzy random variables. METRON-Int. J. Stat. **LXV**, 277–303 (2007)
3. G. Debreu, Integration of correspondences, in *Proceedings of the Fifth Berkeley Symposium of Mathematical Statistics and Probability* (Berkeley, USA, 1965), pp. 351–372
4. A.P. Dempster, Upper and lower probabilities induced by a multi-valued mapping. Ann. Math. Stat. **38**, 325–339 (1967)
5. T. Denœux, Modeling vague beliefs using fuzzy-valued belief structures. Fuzzy Sets Syst. **116**(2), 167–199 (2000)
6. D. Dubois, H. Prade, *Fuzzy Sets and Systems: Theory and Applications* (Academic Press, New York, 1980)
7. D. Dubois, H. Prade, Evidence measures based on fuzzy information. Automatica **21**(5), 547–562 (1985)
8. D. Dubois, H. Prade, *Possibility Theory* (Plenum Press, New York, 1988)
9. R. Féron, Ensembles aléatoires flous. C.R. Acad. Sci. Paris Ser. A **282**, 903–906 (1976)
10. D.G. Kendall, Foundations of a theory of random sets, in *Stochastic Geometry*, ed. by E.F. Harding, D.G. Kendall (Wiley, New York, 1974), pp. 322–376
11. R. Körner, On the variance of fuzzy random variables. Fuzzy Sets Syst. **92**, 83–93 (1997)
12. R. Kruse, On the variance of random sets. J. Math. Anal. Appl. **122**, 469–473 (1987)
13. R. Kruse, K.D. Meyer, *Statistics with Vague Data* (D. Reidel Publishing Company, Dordrecht, 1987)
14. H. Kwakernaak, Fuzzy random variables: definition and theorems. Inf. Sci. **15**, 1–29 (1978)
15. H. Kwakernaak, Fuzzy random variables: algorithms and examples in the discrete case. Inform. Sci. **17**, 253–278 (1979)
16. G. Matheron, *Random Sets and Integral Geometry* (Wiley, New York, 1975)
17. H.T. Nguyen, On random sets and belief functions. J. Math. Anal. Appl. **63**, 531–542 (1978)
18. M. Puri, D. Ralescu, Fuzzy random variables. J. Math. Anal. Appl. **114**, 409–422 (1986)
19. G. Shafer, *A Mathematical Theory of Evidence* (Princeton University Press, Princeton, 1976)
20. G.L.S. Shackle, *Decision, Order and Time in Human Affairs*, 2nd edn. (Cambridge University Press, UK, 1961)
21. C.A.B. Smith, Consistency in statistical inference and decision. J. R. Stat. Soc. B **23**, 1–37 (1961)
22. P. Walley, Measures of uncertainty in expert systems. Artif. Intell. **83**(1), 1–58 (1996)
23. R.R. Yager, A foundation for a theory of possibility. Cybern. Syst. **10**(1–3), 177–204 (1980)
24. R.R. Yager, An introduction to applications of possibility theory. Hum. Syst. Manag. **3**, 246–269 (1983)
25. R.R. Yager, An introduction to applications of possibility theory. Hum. Syst. Manag. **3**, 246–269 (1983)
26. L.A. Zadeh, Fuzzy sets. Inf. Control **8**, 338–353 (1965)
27. L.A. Zadeh, Fuzzy sets as a basis for a theory of possibility. Fuzzy Sets Syst. **1**, 3–28 (1978)
28. L.A. Zadeh, Fuzzy sets and information granularity, in *Advances in Fuzzy Set Theory and Applications*, ed. by M.M. Gupta, R.K. Ragade, R.R. Yager. (North-Holland, Amsterdam, 1979), pp. 3–18

Chapter 2
Random Sets as Ill-Perceived Random Variables

2.1 Introduction: Different Applications of Random Sets

The modern theory of random sets was initiated in the seventies, independently by Kendall [30] and Matheron [37] and it has been fruitfully applied in different fields such as economy, stochastic geometry and ill-observed random objects. Roughly speaking, a random set is a random element in a family of subsets of a certain universe. In particular, we can compute the probabilities that a random set hits a given set, or is included in this set, or yet includes it. In parallel and quite independently of this literature, random sets appear (but for the name) in the late sixties in the pioneering works of Dempster [16], who considers multimappings from a probability space to another space of interest. However Dempster's purpose is to compute the impact of imprecision modeled by the multimapping (that models only a correspondence between the two spaces) on the probability of events in the second space. In particular, the probability that the random set hits the set that models the event of interest, and the probability that it is included in this set, are interpreted as upper and lower probabilities. Dempster's construction is at the origin of the theory of evidence by Shafer [47], that relies on a probability distribution on the power set of a finite set called *frame of discernment*. This theory has given birth to a large literature in Artificial Intelligence and information processing (see Yager and Liu [50] for an anthology). Interestingly there is almost no overlap between the two streams of literature, that is, the Kendall–Matheron view of random sets on the one hand, and the Dempster–Shafer approach, if we expect the classic paper by Nguyen [44] that bridges the gap between the two traditions. In this book, we consider an interpretation of random sets more in line with the Dempster approach, namely we consider them as modeling information tainted with both randomness and incompleteness. This is the topic of this chapter, while the next chapter extends this framework to fuzzy sets.

2.1.1 Two Uses of Sets

The existence of two streams of literature concerning random sets is due to the fact that they do not use the notion of set for the same purpose. As pointed out by Dubois

© The Author(s) 2014   7
I. Couso et al., *Random Sets and Random Fuzzy Sets as Ill-Perceived Random Variables*,
SpringerBriefs in Computational Intelligence, DOI: 10.1007/978-3-319-08611-8_2

and Prade (see [20, 21], for instance) a set can be either used to represent a precise but composite piece of information naturally taking the form of a set (this is called the *conjunctive* or *ontic* view of sets) or as a piece of incomplete information regarding an ill-located element (*disjunctive* or *epistemic* approach).

Under the conjunctive interpretation, a set is regarded as a collection of elements that satisfy some property, or forming some object, or yet representing the precise value of some set-valued attribute. For instance, consider the set of languages that Peter can speak. The piece of information *languages(Peter) = {English, Spanish}* is precise: it means that Peter can speak both English and Spanish, and he is not able to speak any other language. The set is thus the precise representation of an objective entity; this is the *ontic view* of sets.

This interpretation contrasts with the *disjunctive* view, where the set represents incomplete information about a specific point value, and contains all possible (mutually exclusive) values under a given state of knowledge. If the available information about the value is of the form $v \in E$, it means that any value outside the set E is considered impossible, and only one value in E is correct. The set E represents the *epistemic state* of an agent concerning a quantity of interest. Suppose we know that John's height is in the interval [1.65, 1.80], and nothing more. Clearly this piece of information is incomplete as it does not fully inform on the height of John: only one value in [1.65, 1.80] is correct, and this piece of information can be improved later on by acquiring more data on John's height. The latter point is a major difference between the conjunctive and the disjunctive uses of sets. Under the conjunctive (or ontic) view, a set is the best representation of the item of interest, while under the disjunctive (or epistemic) view a set can potentially be reduced into a more precise piece of information, by collecting more knowledge.

2.1.2 Canonical Examples

Here, we illustrate the difference between the two kinds of random sets induced by the different uses of sets, with some typical examples that may help a proper understanding of the point of view adopted in this book. First, we will consider the conjunctive view in Examples 1 and 2.

Example 1 Let us consider the following set of languages

$$L = \{\text{Dutch, English, French, German, Italian, Russian, Spanish}\}.$$

Suppose that we select at random a person from the set Ω_1 of attendees to an international conference, and we ask her/him about the languages (in L) (s)he can speak. We can model the outcome of this experiment by means of a random set that assigns, to each attendee, the collection of languages (s)he can speak. If every attendee to the conference has the same probability to be selected, we will consider the Laplace (uniform) probability distribution P defined on the finite set Ω_1. We can represent this situation by means of a set-valued mapping $\Gamma_1: \Omega_1 \to \wp(L)$, where Ω_1 denotes the population (the set of attendees to the conference) and $\wp(L)$ denotes the family of subsets of L. The triple (Ω_1, P, Γ_1) describes (and quantifies) the variability of the images of the set-valued mapping Γ_1 across Ω_1. We can be interested, for instance, in the proportion of those attendees to the conference that can speak French and English and cannot speak any other language in L. Such a proportion is the probability that the random set Γ_1 takes the "value" {English, French}, i.e., the probability $P(\{\omega \in \Omega_1: \Gamma_1(\omega) = \{\text{English, French}\}\})$ (according to the Laplace distribution on Ω_1) of the set of attendees. Likewise, we could compute the probability that attendees speak at least three languages, which would be $P(\{\omega \in \Omega_1: |\Gamma_1(\omega)| \geq 3\})$.

Example 2 Let us consider the random experiment that consists in randomly selecting a day in a year and recording the interval given by the minimum (lowest recorded) and the maximum (highest recorded) temperature in Oviedo (Spain). We can model this information by means of the Laplace probability defined on Ω_2 (the collection of days in the whole year) and the set-valued mapping $\Gamma_2: \Omega_2 \to \wp(\mathbb{R})$, where $\Gamma_2(\omega) = [T_n(\omega), T_x(\omega)]$ represents the interval of minimum and maximum temperatures attained in Oviedo on a date ω. We aim to study the probability distribution of the interval of minimum and maximum temperatures during the year, in order to calculate the pair of annual average minimum and maximum temperatures, the annual average range (average difference between the maximum and the minimum temperatures), etc.

In both examples, we have considered the outcomes of the respective random sets as "conjunctive" sets: the set-valued outcomes of each of them represents a piece of precise information about a certain attribute of the individuals in the population (attendees or days, respectively). Now, let us exemplify the disjunctive approach.

Example 3 Consider again the set of attendees of an international conference, Ω_1 from Example 1. Imagine that we have imprecise information about the number of languages (within the set of languages L) that each attendee can speak, and that we represent this kind of imprecise information by means of a multi-valued mapping $\Gamma_3: \Omega_1 \to \wp(\{1, 2, \ldots, 7\})$. The random set (Ω_3, P, Γ_3) represents our imprecise information about the random variable $N: \Omega_1 \to \{1, 2, 3, 4, 5, 6, 7\}$ that denotes the number of languages that attendees of the conference can speak. Such multi-valued mapping assigns, to each person, our information about the number of languages (s)he can speak. For instance, if we know that $\omega = $ Didier can speak at least French and English, but we have no additional information regarding his possible knowledge about the rest of the languages in L, then we can represent our imprecise information about the number of languages (in L) he can speak by means of the finite set $\Gamma_3(\omega) = \{2, \ldots, 7\}$. This means that, according to our incomplete information, Didier can speak at least two languages, and no more than seven (the cardinal of the set L) in the list. All we know about the number of languages, $N(\omega)$, that Didier can speak is that such a number belongs to the set $\Gamma_3(\omega)$. Then the probability that attendees can speak, say, two or three languages, will be imprecise, that is lower bounded by $P(\{\omega \in \Omega_3: \Gamma_3(\omega) \subset \{2, 3\}\})$ and upper bounded by $P(\{\omega \in \Omega_3: \Gamma_3(\omega) \cap \{2, 3\} \neq \emptyset\})$.

Example 4 Let us consider again the random experiment from Example 2. But now suppose that we aim to study the distribution of the daily mean temperatures in Oviedo during the year. (The mean temperature of the day will be considered to be the average of all the hourly measurements registered in the day.) Let us assume that, due to storage limitations, we are just provided with the pair of minimum and maximum temperatures, and we do not have any further information about the remaining records during the day. Thus, we know that the mean temperature on a date ω, $T(\omega)$, is greater than or equal to $T_n(\omega)$ and less than or equal to $T_x(\omega)$ (where T_n and T_x denote the random variables considered in Example 2), but we do not know anything else. The random interval $\Gamma_2 = [T_n, T_x]$ represents our imprecise information about the random variable $T \colon \Omega_2 \to \mathbb{R}$.

In Examples 3 and 4, each set-valued outcome represents the ill-perception of a numerical value. In this book, we will focus on this particular application of random sets. Random sets seem to be the natural representation for imprecisely observed random quantities, when we can only determine a set of values that contains each outcome. We will make use of this tool when, due to the imprecision of devices or the incompleteness in experts' information, we can only assure that the outcome belongs to a more or less precise set, but we cannot establish a unique probability distribution within it indicating different grades of belief for different zones of the set.

2.1.3 Encoding Random Sets by Random Vectors

We will now discuss the appropriateness of an alternative formal representation of the information considered in these four examples. In Example 1, the outcomes of the respective random set are finite subsets of the universe L. Thus, we could alternatively encode the information provided by the random set by means of a random binary vector of length 7 (the total number of different languages considered in the example, i.e., the cardinality of L.) For instance, the subset {English, Spanish} could have been alternatively encoded as the vector $(0, 1, 0, 0, 0, 0, 1)$. Analogously, in Examples 2 and 4, the images of the random set Γ_2 are univocally determined by two-dimensional vectors. In fact, the information provided by the random interval $\Gamma_2 = [T_n, T_x]$ can be equivalently represented by means of the two-dimensional random vector $\mathbf{T} = (T_n, T_x)$. Analogously, the information provided in Example 3, can be equivalently represented by means of a two-dimensional random vector. In fact, the outcomes of the random set Γ_3 are subsets of the finite set of integers $\{1, \ldots, 7\}$ determined by the pair of their minimum and maximum values. Therefore, Γ_3 can be alternatively encoded by means of the two-dimensional random vector $\mathbf{N} = (N_n, N_x)$, where $N_n(\omega)$ and $N_x(\omega)$ respectively denote the minimum and the maximum of the finite set $\Gamma_3(\omega)$. (In this special case, both components of the random vector take values in the finite set of numbers $\{1, \ldots, 7\}$.) For instance, the outcome associated to Didier would be the pair of numbers $(2, 7)$, according to this random-vector representation.

Summarizing, in all the above examples, the set-valued images of the respective random sets are univocally determined by means of finite tuples of values, and therefore all those random sets can be equivalently expressed by means of random vectors. Up to this point, one might wonder whether we really need to use random sets in order to manage the information provided in the above examples, or not. In fact, it is not difficult to check that the probability measure induced by each of the above random sets (when considered as a measurable function on a σ-algebra of families of sets) is univocally determined by the joint probability measure induced by the associated random vector.

Notwithstanding, this kind of vector-valued information is not enough in the disjunctive framework. The probability distribution on vectors induced by the random set does not contain all the information conveyed by the random set about the probability measure induced by the underlying ill-known variable. In fact, the (imprecise) information that the random set provides about the probability distribution induced by the *original* ill-perceived random variable is not only captured by the probability measure induced by the random set, but it also depends on the nature of the initial space where it is defined, as we will show in the forthcoming sections. Thus, we will need some tools from the theory of Imprecise Probabilities [49] in order to describe this type of imprecise information about the *original probability distribution*. Regarding the conjunctive framework, referring to such a theory is not necessary in general, and the random set probability distribution is enough to represent all the relevant information. Therefore, in order to manage situations such as those described in Examples 1 and 2, classical statistical tools applied to random vectors would be sufficient. This is not the case in every application of random sets under the "conjunctive" approach. In fact, in many problems arising, for instance, in the frameworks of Economy or Spatial Statistics, the shape of the set-valued random images is not so simple (they are not, in general, subsets of a finite universe or intervals in the real line), and therefore we cannot find a random vector representation that captures all the information provided by the random set. Anyway, this kind of problems falls into another area, called Random Set Theory, distinct from the Dempster-Shafer tradition, and that is out of the scope of this book.

2.2 Basic Formal Definitions

The earliest works on random sets date from the 1930s, with the pioneer works by Novikov [45] and Lusin [35]. They were later studied by many different authors such as Choquet [5], Aumann [1], Castaing and Valadier [3], Debreu [15] or Hildenbrand [27] in different contexts, such as Stochastic Geometry, Economics or Convex Analysis. The interested reader can furthermore consult the monographs by Matheron [37], Molchanov [42, 43] or Stoyan et al. [48] for an overall view of the formal aspects and the main applications of random sets under the "conjunctive" approach. In this section, we will not provide a general, nor an exhaustive introduction to the Theory of Random Sets, but we will restrict ourselves to those basic formal notions needed under the "disjunctive" approach, the subject of the rest of the book, in the spirit of Dempster [16].

A random set is a multi-valued mapping defined on a probability space that satisfies some measurability condition with respect to some σ-algebra defined on a family of subsets of the final space. The different definitions in the literature disagree on the measurability conditions imposed to this mapping, and in the properties of the output space. In this book, we will assume the multi-valued mapping to be *strongly measurable* [44]. In order to introduce such a definition, we do not need to assume any special topological structure on the final space. In fact, we just need it to be a measurable space.

2.2.1 Upper and Lower Inverses

The concept of strong measurability is based on the notions of *upper* and *lower inverse* provided below.

Definition 1 [15, 16, 44] Let us consider a probability space, $(\Omega, \mathcal{A}, P)$, a measurable space $(U, \mathcal{A}')$ and a multi-valued mapping $\Gamma \colon \Omega \to \mathcal{P}(U)$ with non-empty images. For $A \subseteq U$, we define the *upper inverse* as

$$\Gamma^*(A) := \{\omega \in \Omega \mid \Gamma(\omega) \cap A \neq \emptyset\},$$

and the *lower inverse* of Γ as

$$\Gamma_*(A) = \{\omega \in \Omega \mid \Gamma(\omega) \subseteq A, \Gamma(\omega) \neq \emptyset\}.$$

According to the above definition, the upper inverse of A is the set of elements in Ω whose images "touch" or "hit" the subset A of U. Furthermore, the lower inverse is the set of elements of Ω whose images are non-empty and are included in A. Therefore it is included in the upper inverse, by definition ($\Gamma_*(A) \subseteq \Gamma^*(A)$). Thus, if we assume that the multi-valued images are non-empty, then the lower inverse of A can also be seen as the set of elements in Ω whose images do not "hit" the complement of A. According to this last assertion, we can easily notice that, when the multi-valued mapping has non-empty images, the lower inverse of A coincides with the complement of the upper inverse of A^c, where A^c denotes the complement of A, $A^c = U \setminus A$. Mathematically:

$$\Gamma_*(A) = [\Gamma^*(A^c)]^c.$$

Furthermore, the notions of upper and lower inverse generalize that of the *inverse image* of events under a random variable. Suppose that the images of Γ are singletons. Then, there exist a point-valued function $X \colon \Omega \to U$ such that $\Gamma(\omega) = \{X(\omega)\}, \forall \omega \in \Omega$. Under this condition, the lower and the upper inverse of any event A do coincide with the inverse image of A under X, as we check below:

$$\Gamma^*(A) = \{\omega \in \Omega : \{X(\omega)\} \cap A \neq \emptyset\} = \{\omega \in \Omega : X(\omega) \in A\} = X^{-1}(A) \quad (2.1)$$

$$\Gamma_*(A) = \{\omega \in \Omega : \{X(\omega)\} \subseteq A, \{X(\omega)\} \neq \emptyset\} = \{\omega \in \Omega : X(\omega) \in A\} = X^{-1}(A).$$
$$(2.2)$$

Having introduced the notions of upper and lower inverse of an event, we can recall the notion of *strong measurability* of a multi-valued mapping.

Definition 2 [44] Let us consider a probability space $(\Omega, \mathcal{A}, P)$, a measurable space $(U, \mathcal{A}')$, and a multi-valued mapping $\Gamma : \Omega \to \mathcal{P}(U)$ with non-empty images. Γ is said to be *strongly measurable* when $\Gamma^*(A) \in \mathcal{A}$, for all $A \in \mathcal{A}'$.

According to duality relation

$$\Gamma_*(A) = [\Gamma^*(A^c)]^c,$$

we can straightforwardly check that Γ is strongly measurable if and only if the lower inverse, $\Gamma_*(A)$, of any measurable subset $A \in \mathcal{A}'$ belongs to $\mathcal{A}$. When there is no risk of confusion about what is the multi-valued mapping under consideration, we will use the nomenclature $A^* = \Gamma^*(A)$ and $A_* = \Gamma_*(A)$, $\forall A \in \mathcal{A}$, for the sake of simplicity.

2.2.2 Random Sets

Definition 3 Let us consider a probability space $(\Omega, \mathcal{A}, P)$, a measurable space, $(U, \mathcal{A}')$. A random set is a strongly measurable multimapping $\Gamma : \Omega \to \mathcal{P}(U)$ with non-empty images.

Let us now check that a random set, viewed as an $\mathcal{A} - \mathcal{A}'$ strongly-measurable multi-valued mapping can be also seen as measurable function in the classical sense. Let us first consider, for each $A \in \mathcal{A}'$, the family of sets:

$$\mathcal{F}_A := \{B \subseteq U \mid B \cap A \neq \emptyset\}$$

and let $\mathcal{C}$ be the collection of all those families, when A ranges over the whole σ-algebra $\mathcal{A}'$, i.e.,

$$\mathcal{C} = \{\mathcal{F}_A \mid A \in \mathcal{A}'\}.$$

Now let us consider the σ-algebra generated by such a collection of families of sets. Let the reader notice that the sets $B \subseteq U$ are "elements" of the power set $\wp(U)$, and the families of the form $\mathcal{F}_A$, therefore, are subsets of elements in $\wp(U)$. The σ-algebra $\sigma(\mathcal{C})$ generated by $\mathcal{C}$ is the least σ-algebra that contains such a collection. We can easily check that a multi-valued mapping from Ω to U is strongly measurable if and only if it is $\mathcal{A} - \sigma(\mathcal{C})$ measurable when considered as a "point-valued" mapping from Ω to $\wp(U)$. This fact will be important in the following sections, when we refer to the probability measure induced by the random set on the σ-algebra $\sigma(\mathcal{C})$.

We shall denote by P_Γ this probability measure induced by Γ on $\sigma(\mathcal{C})$ (when considered as a "point-valued" function). It assigns to any element $\mathcal{S}$ of $\sigma(\mathcal{C})$ the probability $P_\Gamma(\mathcal{S})$. When U is finite, the random set Γ induces a probability distribution denoted by m over the power set of U, such that $m(E) = P(\{\omega \in \Omega : \Gamma(\omega) = E\})$, called basic mass assignment (bma). It is clear that $\sum_{A \subseteq U} m(E) = 1$, and (if Γ has no empty image) $m(\emptyset) = 0$. Then, $\sigma(\mathcal{C}) = \wp(U)$ and $\forall \mathcal{S} \subseteq \wp(U), P_\Gamma(\mathcal{S}) = \sum_{A \in \mathcal{S}} m(A)$.

The notion of strong measurability does not require any special topological structure on the final space. Other definitions of measurability of multi-valued mappings in the literature assume some additional structure. All of them generalize the notion of measurability of a point-valued function, in the sense that all of them are equivalent to each other when the images of the multi-valued mapping are singletons. The formal relations between different measurability conditions under different frameworks have been extensively studied in the literature. We will not focus on these formal aspects. The interested reader can consult [15, 25, 28, 44] for further details. Throughout this book, we will use term *random set* to refer to a strongly-measurable multi-valued function.

In the following sections, the random set Γ is considered as the representation of an ill-perceived random variable and we shall clarify the role that the probability measure P_Γ on $\sigma(\mathcal{C})$ plays under the epistemic interpretation.

2.2.3 Upper and Lower Probabilities

Based on the notions of upper and lower inverse of an event, and the definition of strong measurability defined above, we can recall the notions of upper and lower probabilities of an event $A \in \mathcal{A}'$.

Definition 4 [16] Let us consider a probability space $(\Omega, \mathcal{A}, P)$, a measurable space, $(U, \mathcal{A}')$, and a random set $\Gamma : \Omega \to \mathcal{P}(U)$, with non-empty images. We define the *upper probability* of $A \in \mathcal{A}'$ as the probability of its upper inverse, i.e.:

$$P_\Gamma^*(A) := P(\Gamma^*(A)) = P(A^*). \tag{2.3}$$

Analogously, the *lower probability* of A is defined as the probability of the lower inverse, i.e.:

$$P_{*\Gamma}(A) := P(\Gamma_*(A)) = P(A_*). \tag{2.4}$$

The above definitions only make sense when Γ is strongly measurable, otherwise the lower and upper inverses of the "events" $A \in \mathcal{A}'$ do not necessarily belong to the initial σ-algebra $\mathcal{A}$, and therefore they might not be considered as arguments of the probability measure P. When there is no confusion about the multi-valued mapping under study, we will use the simplified nomenclature $P^*: = P_\Gamma^*$ and $P_*: = P_{*\Gamma}$. According to the definition of upper and lower image of an event A, the upper probability, $P^*(A)$ denotes the probability that the outcome of the random set "hits" the set A. Therefore, the set function $P^*: \mathcal{A}' \rightarrow [0, 1]$ is also called the *hitting function*.

Analogously, the lower probability of A, $P_*(A)$, represents the probability that the outcome of Γ is included in A, or equivalently, the probability that it "misses" A^c. Taking into account the relation $A_* = [(A^c)^*]^c$, we can easily check that P^* and P_* satisfy the property of duality:

$$P_*(A) = 1 - P^*(A^c), \quad \forall A \in \mathcal{A}',$$

so they are *conjugate* to each other. Therefore, each of them univocally determines the other one.

The set functions P_* and P^* are non-additive, in general. In fact, they only satisfy the additivity property when the images of the random set are singletons almost everywhere. In the general case, they just satisfy the following (less restrictive) inequalities: $\forall, k \in \mathbb{N}$:

$$P^*\left(\bigcap_{i=1}^{k} A_i\right) \le \sum_{i=1}^{k} P^*(A_i) - \sum_{i<j} P^*(A_i \cup A_j) + \cdots + (-1)^{k+1} P^*(A_1 \cup \ldots \cup A_k); \quad (2.5)$$

$$P_*\left(\bigcup_{i=1}^{k} A_i\right) \ge \sum_{i=1}^{k} P_*(A_i) - \sum_{i<j} P_*(A_i \cap A_j) + \cdots + (-1)^{k+1} P_*(A_1 \cap \ldots \cap A_k). \quad (2.6)$$

According to the above equations, and following Choquet's nomenclature [5, 17], they satisfy the properties of ∞-alternating and an ∞-monotone capacities, respectively.

As we have noticed above, when, in particular, the images of the random set are singletons, both set functions do coincide and they satisfy the additivity property. Thus, the notions of upper and lower probability generalize the definition of probability measure induced by a random variable. In fact, in this particular case, there exists a point-valued mapping $X: \Omega \rightarrow U$ such that $\Gamma(\omega) = \{X(\omega)\}$, $\forall \omega \in \Omega$, and the upper and lower images of any $A \in \mathcal{A}'$ under Γ do satisfy the equalities:

$$\Gamma^*(A) = \Gamma_*(A) = X^{-1}(A)$$

as we have checked in Eqs. 2.1 and 2.2. According to these equalities, the above mapping X is $\mathcal{A} - \mathcal{A}'$ measurable whenever the multi-valued mapping $\Gamma = \{X\}$ is strongly measurable. Furthermore, in this particular situation, the following equalities hold:

$$P(\Gamma^*(A)) = P(X^{-1}(A)) = P(\Gamma_*(A))$$

or, equivalently,

$$P^*(A) = P_X(A) = P_*(A).$$

Remark 1 Another special case is when the set $\{\Gamma(\omega): \omega \in \Omega\}$ of set-valued images of Γ is nested. Then, the set functions P^* and P_* are generally[1] respectively sup-preserving and inf-preserving [6], in particular $P^*(A \cup B) = \max(P^*(A), P^*(B))$ and $P_*(A \cap B) = \min(P^*(A), P^*(B))$. In this situation, P^* and P_* are said to be consonant, and coincide with possibility and necessity measures, respectively [18, 19]. Then the information is entirely contained in the possibility distribution on U defined by $\pi(u) = P(\Gamma^*(\{u\})$, namely $P^*(A) = \sup_{u \in A} \pi(u)$. This concept will be instrumental when discussing the epistemic view of random fuzzy sets. See Sect. 3.1.1 in the next chapter.

For an arbitrary random set Γ, the set functions P^* and P_* are determined by the probability measure P_Γ induced by the random set Γ on the σ-algebra $\sigma(\mathcal{C})$ introduced in Sect. 2.2.2: For all $A \in \mathcal{A}'$, the upper inverse A^* can be seen as the set of elements in Ω whose images belong to the family of (hitting) sets $\mathcal{F}_A$. Indeed, we observe that $P^*(A) = P_\Gamma(\mathcal{F}_A)$. Similarly, we can observe that A_* is the set of values whose images under Γ do not belong to $\mathcal{F}_{A^c}$. Therefore, we have that $P_*(A) = 1 - P_\Gamma(\mathcal{F}_{A^c})$. Conversely, the upper and lower probabilities of Γ also determine the probability measure induced by the random set Γ on $\sigma(\mathcal{C})$, since the class of families of sets $\{\mathcal{F}_A^c: A \in \mathcal{A}'\}$ is closed under intersections. This is part of the well-known Choquet Theorem. This fact will be important in the following sections, where the random set Γ is considered as the representation of an ill-perceived random variable, in order to clarify the role that the probability measure induced by Γ on $\sigma(\mathcal{C})$ plays in the epistemic setting.

The upper and lower probabilities induced by Γ on U, as well as the probability measure induced by Γ, when considered as a "point-valued" function on $\wp(U)$ are also univocally determined by the *credal set* of Γ:

Definition 5 The *credal set* induced by Γ is the convex family of probability measures that are *dominated* by its upper probability, P^*:

$$\mathcal{M}(P^*) := \{P: \mathcal{A}' \to [0, 1] \text{ such that } P(A) \leq P^*(A), \quad \forall A \in \mathcal{A}'\}.$$

According to the duality relation between P^* and P_* the credal set can be alternatively expressed as:

$$\mathcal{M}(P^*) := \{P: \mathcal{A}' \to [0, 1]: P(A) \geq P_*(A), \quad \forall A \in \mathcal{A}'\}.$$

---

[1] The interested reader can consult Refs. [6, 38] for detailed formulations of sufficient conditions. Those conditions include the very common situations where the images of Γ are compact subsets of $\mathbb{R}^n$ or the final space is finite.

In the next section, we will consider random sets as representations of imprecise observations of random variables. We will make use of all the above formal notions in order to clarify how to deal with this kind of incomplete information.

Remark 2 When U is finite, the random set Γ induces a probability distribution m over the power set of U, such that $m(E) = P(\{\omega \in \Omega : \Gamma(\omega) = E\})$, called basic mass assignment (bma). As we noticed in Sect. 2.2.1, $\sum_{A \subseteq U} m(E) = 1$, and (if Γ has no empty image) $m(\emptyset) = 0$. In his theory of evidence, Shafer [47] considers the bma as the primitive information representing the knowledge of an agent, thus doing away with the underlying probability space. Then $m(A)$ is interpreted as the subjective probability that the agent only knows that some ill-known value $X_0 \in A$, the lower probability models how much the agent believes in A (and is then denoted by $Bel(A) = \sum_{E \subseteq A} m(E)$), while the upper probability models how much the agent finds A plausible (and is then denoted by $Pl(A) = \sum_{E \cap A \neq \emptyset} m(E)$). See also Exercise 6.

2.3 Three Sets of Probabilities Associated with a Random Set

In Sect. 2.2, we have recalled the notions of upper and lower probabilities induced by a random set. When, in particular, the images of the random set are singletons, both of them coincide with the notion of probability measure induced by a random variable. We will show in this section that, in the general case, when the random set represents incomplete information about an ill-observed random variable X_0, its upper and lower probability bounds (respectively above and below) enclose the "true" probability measure[2] induced by X_0. i.e.:

$$P_*(A) \leq P_{X_0}(A) \leq P^*(A), \quad \forall A \in \mathcal{A}'$$

In other words, the probability measure induced by X_0 on $\mathcal{A}'$ is included in the credal set $\mathcal{M}(P^*)$. Furthermore, as we have noticed in Sect. 2.2 such a credal set is univocally determined by the probability measure P_Γ induced by the random set (viewed as a measurable random object). Therefore, P_Γ, provides some information about the probability measure P_{X_0} but, as we will check in the next section, it does not capture all the information that the random set contains about it. In fact, we will check that the information provided by Γ about such a probability measure is determined by the "probability envelope" of Γ, that is, a family of probability measures included in the credal set, and sometimes a proper subset of it. Such a family is not determined in general, by the probability measure induced by Γ, as we will show with the help of some illustrative examples.

Let us suppose that a random set $\Gamma : \Omega \to \wp(U)$ represents the imprecise measurements of the outcomes of an otherwise point-valued random variable

---

[2] In Shafer evidence theory, this true probability is not supposed to exist, as $Bel(A)$ *is* the degree of belief in A.

$X_0: \Omega \to U$. In other words, let us assume that the random set Γ restricts mutually exclusive possible outcomes (according to the "disjunctive" interpretation of sets) and therefore the image $\Gamma(\omega)$ denotes the set of possible values for the outcome of ω under the random variable X_0: for each $\omega \in \Omega$, all we know about $X_0(\omega)$ is that it belongs to the set $\Gamma(\omega)$. According to this setting, the random set is regarded as a collection of feasible random variables. Such collection is called the class of *measurable selections* of Γ:

Definition 6 Consider two measurable spaces $(\Omega, \mathcal{A})$ and $(U, \mathcal{A}')$. A function $X: \Omega \to U$ is said to be a *measurable selection* of a multi-valued mapping $\Gamma: \Omega \to \wp(U)$ when:

- Its images are contained in those of Γ, i.e., $X(\omega) \in \Gamma(\omega),\ \forall \omega \in \Omega$.
- It is $\mathcal{A} - \mathcal{A}'$ measurable.

We will denote by $S(\Gamma)$ the family of all the measurable selections of Γ:

$$S(\Gamma) = \{X: \Omega \to U,\ \mathcal{A} - \mathcal{A}'\text{—measurable}: X(\omega) \in \Gamma(\omega),\ \forall \omega \in \Omega\}.$$

According to the above information, all we know about the probability measure induced by the random variable X, P_X, is that it belongs to the family of probability measures induced by the measurable selections of Γ. We will call it the *probability envelope* of Γ:

Definition 7 Consider a probability space $(\Omega, \mathcal{A}, P)$, a measurable space $(U, \mathcal{A}')$ and a multi-valued mapping $\Gamma: \Omega \to \wp(U)$. We will call the following class of probability measures:
$$\mathcal{P}(\Gamma) = \{P_X: X \in S(\Gamma)\},$$

the *probability envelope* of Γ.

We can easily check that the class $\mathcal{P}(\Gamma)$ is included in the credal set $\mathcal{M}(P^*)$. In fact, for any measurable selection $X \in S(\Gamma)$, and an arbitrary event $A \in \mathcal{A}'$, the inverse image $X^{-1}(A)$ is included in the upper-image A^*. (We assume that $X(\omega) \in \Gamma(\omega), \forall \omega \in \Omega$, and thus we easily observe that $X(\omega) \in A$ implies $\Gamma(\omega) \cap A \neq \emptyset$.) Therefore the following inequality holds:

$$P_X(A) = P(X^{-1}(A)) \leq P(A^*) = P^*(A),\quad \forall A \in \mathcal{A}'.$$

Thus, for any measurable selection X, the induced probability measure P_X is dominated by P^* and therefore it belongs to the credal set $\mathcal{M}(P^*)$. In other words, the probability envelope $\mathcal{P}(\Gamma)$ is always included in the credal set $\mathcal{M}(P^*)$. Moreover, such an inclusion can be strict in very common situations, as we will illustrate in Example 5. The credal set $\mathcal{M}(P^*)$ is univocally determined by the probability measure induced by Γ on $\sigma(\mathcal{C})$, as we recalled in the last section. But, the probability envelope $\mathcal{P}(\Gamma)$ is not, in general. In fact, there can be found two different random sets with the same probability distribution, but with different probability envelopes, as we will see in Example 6. This means that the probability measure induced by Γ, when considered as a measurable "point-valued" mapping on $\wp(U)$, does not keep all the information that Γ provides about the probability distribution induced by X_0 on $\mathcal{A}'$. This issue plays a central role within the treatment of ill-observations of random variables, because it forces us to use "non-classical" probability techniques. In this section, we will show some examples to illustrate this fact.

Let us start with an example where the probability envelope of a random set is strictly included in the credal set and, furthermore, is not determined by the probability measure induced by the random set.

Example 5 Let us consider a singleton $\{\omega_0\}$ endowed with the structure of a probability space with the σ-algebra $\wp(\{\omega_0\}) = \{\emptyset, \{\omega_0\}\}$ and the probability measure $P: \wp(\{\omega_0\}) \to [0, 1]$ defined as $P(\emptyset) = 0$, and $P(\{\omega_0\}) = 1$. Let us also consider the multi-valued mapping $\Gamma: \{\omega_0\} \to \wp(\mathbb{R})$ defined as $\Gamma(\omega_0) = [a, b]$, where a and b are two arbitrary values satisfying the restriction $a < b$. We straightforwardly check that Γ is strongly measurable with respect to any σ-algebra defined on the real line, as the upper inverse of any subset of $\mathbb{R}$ belongs to the σ-algebra $\wp(\{\omega_0\})$. So, if we consider the usual Borel σ-algebra on $\mathbb{R}$, $\beta_{\mathbb{R}}$, the upper probability associated to Γ is the set function $P^*: \beta_{\mathbb{R}} \to [0, 1]$ defined as follows:

$$P^*(A) = P(\{\omega \in \{\omega_0\}: \Gamma(\omega)\} \cap A \neq \emptyset).$$

According to the definition of the random set Γ, the set function $P^*: \beta_{\mathbb{R}} \to [0, 1]$ can be expressed as a Boolean possibility measure [19]:

$$P^*(A) = \begin{cases} 1 & \text{if } A \cap [a, b] \neq \emptyset \\ 0 & \text{otherwise} \end{cases}$$

The credal set $\mathcal{M}(P^*)$ is the set of all those probability measures dominated by P^*. Therefore it is the set of probability measures that assign null probability to all those measurable sets that do not intersect the interval $[a, b]$:

$$\mathcal{M}(P^*) = \{P: \beta_{\mathbb{R}} \to [0, 1] \ \text{ such that } \ [A \cap [a, b] = \emptyset \Rightarrow P(A) = 0]\}.$$

In other words, it is the set of probability measures whose support is included in the interval $[a, b]$:

$$\mathcal{M}(P^*) = \{P: \beta_{\mathbb{R}} \to [0, 1] \ \text{ such that } \ P([a, b]) = 1\}.$$

Besides, the set of measurable selections of Γ is the set of all those point valued mappings defined on $\{\omega_0\}$ whose image belongs to the interval $[a, b]$:

$$S(\Gamma) = \{X: \{\omega_0\} \to \mathbb{R}: X(\omega_0) \in [a, b]\}.$$

Since the initial space is a singleton, any of those measurable selections is associated to a deterministic experiment. Therefore, the probability envelope of Γ is the set of all degenerated probability measures with support included $[a, b]$:

$$\mathcal{P}(\Gamma) = \{\delta_x : x \in [a, b]\},$$

where δ_x denotes the Dirac measure on x ($\delta_x(A) = 1_A(x), \forall x \in \mathbb{R}, \forall A \in \beta_\mathbb{R}$, where 1_A denotes the indicator function of A.) Thus, we observe that $\mathcal{P}(\Gamma)$ is strictly included in the credal set $\mathcal{M}(P^*)$.

Furthermore, we can observe that the probability envelope $\mathcal{P}(\Gamma)$ cannot be expressed in general as a function of the probability distribution of Γ, when considered as a "classical" measurable function. In fact, we can define two different random sets inducing the same probability measure, but different probability envelopes as we will next illustrate.

Example 6 Let us consider the probability space $([0, 1], \beta_{[0,1]}, \lambda_{[0,1]})$ that represents the unit interval endowed with the Lebesgue measure (the uniform distribution) and let $\Gamma' : [0, 1] \to \wp(\mathbb{R})$ be the constant multi-valued mapping defined on it as $\Gamma'(\omega) = [a, b], \forall \omega \in [0, 1]$. We can easily check that the upper inverse of any event $A \in \beta_\mathbb{R}$ coincides either with the empty set or with the whole unit interval, and, therefore, the multi-valued mapping Γ' is $\beta_{[0,1]} - \beta_\mathbb{R}$—strongly measurable.

Furthermore, it induces the same upper probability as the random set $\Gamma :$ $\{\omega_0\} \to \wp(\mathbb{R})$ defined in Example 5. In fact, both of them take the "value" $[a, b]$ with probability equal to 1. Notwithstanding, the probability envelope of Γ' does not coincide with that of Γ. In fact, the set of measurable selections of Γ' is the set:

$$S(\Gamma') = \{X : [0, 1] \to [a, b] : X \text{ is } \beta_{[0,1]} - \beta_\mathbb{R} \text{ measurable and } X(\omega) \in [a, b], \forall \omega \in [0, 1]\},$$

and therefore, the probability envelope of Γ' is the set of all probability measures induced by all the random variables in the set $S(\Gamma')$. It is well known that, for any probability measure, there exists a random variable defined on the unit interval endowed with the Lebesgue measure that induces it. According to it, the probability envelope of Γ' is the family

$$\mathcal{P}(\Gamma') = \{P : \beta_\mathbb{R} \to [0, 1] \text{ such that } P([a, b]) = 1\},$$

which, by the way, coincides with the credal set $\mathcal{M}(P^*)$.

Example 5 is somehow artificial, since there is not any random experiment of real interest where the population under study is a singleton. Notwithstanding, it helps us clarify why the probability envelope of a random set is not determined, in general, by its induced probability distribution. Each of the random sets considered in the last examples is defined on a different initial space: In the first case (Example 5), the initial space just contains one atom and therefore the probability envelope consists of a set of degenerate probability distributions. On the other hand, in the second case (Example 6), Γ' is defined on a non-atomic space (the unit interval endowed with the Lebesgue measure) and therefore the probability envelope contains all the probability measures with support in the interval $[a, b]$. Informally speaking, the "less atomic" the initial space, the wider the probability envelope. This is an important issue when we deal with imprecise observations of variables defined on finite populations. According to the above reasoning, the "size" of the probability envelope (that represents the available information about the probability measure associated to the ill-observed random variable) is very closely related to the size of the population.

The formal relationships between the probability envelope and the credal set associated to a random set have been studied in detail in [4, 39–41], among others. It is proved in [41] that the supremum and the infimum of the probability envelope:

$$\overline{P}_\Gamma(A) = \sup_{Q \in \mathcal{P}(\Gamma)} Q(A), \quad \underline{P}_\Gamma(A) = \inf_{Q \in \mathcal{P}(\Gamma)} Q(A)$$

are indeed a maximum and a minimum. On the other hand, this pair of maximum and minimum set-valued mappings does not coincide respectively with P^* and P_*, in general, although there have been found some sufficient conditions in [41]. Those sufficient conditions include, as particular cases, the very common situation when the final space U is finite, or where the images of Γ are compact subsets of $\mathbb{R}^n$.

But, even in those cases where P^* and P_* do respectively coincide with the maximum and the minimum of $\mathcal{P}(\Gamma)$, they do not necessarily determine the probability envelope $\mathcal{P}(\Gamma)$. (A detailed formal study regarding this issue can be found in [39] and [40], for instance.)

Actually there is a third set of probability measures that can be associated to an epistemic random set. Namely we can compute for each event $A \in \mathcal{A}'$ the set of possible probability values induced by Γ as

$$\mathcal{P}_\Gamma(A) = \{P_X(A): X \in \Gamma\} = \{Q(A): Q \in \mathcal{P}(\Gamma)\}.$$

Now consider the set

$$\mathcal{P}^\downarrow(\Gamma) = \{Q: Q(A) \in \mathcal{P}_\Gamma(A), \forall A \in \mathcal{A}'\}.$$

It should be clear that

$$\mathcal{P}(\Gamma) \subseteq \mathcal{P}^\downarrow(\Gamma) \subseteq \mathcal{M}(P^*).$$

In general, these sets will not coincide. For one, $\mathcal{P}(\Gamma)$ and $\mathcal{P}^\downarrow(\Gamma)$ are generally not convex (e.g. if Ω is finite, they are both finite too). Moreover, the sets $\mathcal{P}_\Gamma(A), \forall A \in \mathcal{A}'$ are projections of $\mathcal{P}(\Gamma)$ on events, and $\mathcal{P}^\downarrow(\Gamma)$ loses information with respect to $\mathcal{P}(\Gamma)$ as the former is built from these projections.

This state of facts has some relevant implications, since even in those cases where the probability distribution of the random set determines the most committed bounds for the probability of any event in the final space, the differences between $\mathcal{P}(\Gamma)$ and $\mathcal{M}(P^*)$ may impact the calculation of other parameters. This will be the main topic in the next subsection (see also examples of consequences of these results in the last part of the next chapter).

2.4 Expectation and Variance of Random Sets

The probability measure induced by the random set (or equivalently, the upper probability P^*, or the credal set $\mathcal{M}(P^*)$) does not provide enough information in order to calculate the most committed bounds for some characteristic parameters of the ill-known probability distribution, such as the variance or the entropy. We will illustrate this issue in forthcoming examples.

2.4.1 The Conjunctive Versus the Disjunctive Approach

First we need to clarify the notion of "characteristic parameter" in our context. We can find in the random set literature at least two alternative procedures to extend it:

(A) According to the conjunctive approach, the random set is viewed as a "random object", i.e., a particular measurable mapping within the framework of classical probability theory (a measurable mapping in a classical sense). So, it induces a probability measure on the σ-algebra $\sigma(\mathcal{C})$. Each parameter is calculated as a function of such a probability measure. More specifically, set-valued arithmetic is used to derive a method of construction of the expectation: a limit-based construction analogous to Lebesgue integral definition (using Minkowski sum and scalar product to define the expectation of "simple" random sets) leads to a definition of expectation which is consistent with Bochner [2] integral.[3] This expectation is a subset of the final space and it plays the role of the average "value" of the random set. We can devise a parallel construction for the variance: let us consider a particular metric, d, defined over the class of subsets of the final space. In this setting, we can define the variance of a random set as the mean (classical expectation of a random variable) of the squared distances from the images of the random set to the (set-valued) expectation:

$$\text{Var}(\Gamma) = \int d^2(\Gamma(\omega), E(\Gamma)) \, dP(\omega).$$

---

[3] The notion of Bochner integral extends the definition of Lebesgue integral to random "objects" that take values in a Banach space, following a similar scheme. First, the integral of a simple random set is defined as the weighted sum of its images (according to the addition operator defined on the Banach space). Finally, the integral of a more general function is defined as the limit of the integrals of appropriate simple functions.

For instance the definitions by Feng et al. [23], Körner [31] and Lubiano [34] fit this intuition. Different definitions in the literature disagree on the particular choice of the metric d. In this context the variance is a (precise) number that quantifies the degree of dispersion of the images of the random set.

(B) The disjunctive approach regards the random set as a collection of random variables (the class of its measurable selections.) The parameter is calculated as the set of "admissible" values. Aumann expectation [1] and Kruse variance [32], for instance, fit this formulation. The sample mean and variance are calculated in a similar way, as sets of admissible values for the mean and variance of the underlying imprecisely observed data. Even when it is defined as a set of admissible values, Aumann expectation is sometimes considered in the literature within the first approach [option (A)], because, under some conditions about the non-atomicity of the initial space, it can be written as a function of the probability distribution of the random set. When the random set is regarded as the representation of the ill-perception of an otherwise point-valued random variable, we must follow this second option. Let us suppose that $\Gamma: \Omega \to \wp(\mathbb{R})$ represents some incomplete knowledge about a point-valued mapping $X_0: \Omega \to \mathbb{R}$. In other words, all we know about each outcome $X_0(\omega)$ is that it belongs to the set $\Gamma(\omega)$. Then, all we know about X_0 is that it is a measurable selection of Γ. Consider an arbitrary parameter $\theta(X_0)$ associated to the probability distribution of X_0. The random set Γ provides the following information about $\theta(X_0)$:

$$\theta(\Gamma) = \{\theta(X): X \in S(\Gamma)\} \tag{2.7}$$

as suggested by Kruse and Meyer in [33]. In other words, if we observe $\Gamma(\omega)$ as an incomplete perception of $X_0(\omega)$, for each $\omega \in \Omega$, then all we know about $\theta(X_0)$ is that it belongs to the class $\theta(\Gamma)$. When, in particular, $\theta(X_0)$ represents the Lebesgue expectation, $E(X_0)$, the collection of values $\theta(\Gamma)$ is called the *Aumann expectation*[4] [1] and, when it represents the variance, $\theta(\Gamma)$ is called the *Kruse variance* [32]. This formulation can be as well applied to entropy (computing the class of entropy values associated to the measurable selections of Γ). The imprecise distribution function studied in [12] also fits this formulation.

2.4.2 Examples

Let us compare both approaches with the help of several illustrative examples.

Example 7 Let $\Gamma = [T_n, T_x]$ be the random set considered in Example 2. The pair of temperatures, $(T_n(\omega), T_x(\omega))$, registered on a particular day ω is considered as an "entity" and not as a pair of bounds for an ill-known numerical value. Under this

---

[4] In fact, the set of all integrable selections, $S^1(\Gamma) \subseteq S(\Gamma)$, instead of the whole class $S(\Gamma)$ is considered here.

approach, it makes sense to calculate the pair of annual minimum and maximum expected temperatures, as well as the dispersion of the outcomes of Γ, regarded as a random object. According to this approach, the interval that ranges from the expected minimum temperature to the expected maximum temperature can be considered as the "average" min-max temperature interval: $E(\Gamma) = [E(T_n), E(T_x)]$. Analogously, the "dispersion" of the outcomes of the random set Γ can be calculated as the expected squared distance between the observed intervals $[T_n(\omega), T_x(\omega)]$ and the interval-valued expectation, $E(\Gamma) = [E(T_n), E(T_x)]$, according to a specific distance function defined on the family of pairs of closed intervals. The specific value of the variance will depend on the selected metric. Two of the most popular definitions of scalar variances are those introduced by Körner [31] and Lubiano [34]. Both definitions of scalar variances are based on both specific families of metrics. In both cases, the variance can be regarded as a function of the mid-point, $\frac{T_n+T_x}{2}$ and the amplitude, $T_x - T_n$, of the outcomes of the random set Γ. Each particular metric will derive a particular function. But, regardless of the choice of the metric, such a function will be increasing in both components. In words, the greater the dispersion, along the year, of the daily temperature ranges and of the daily half-sum temperatures, the greater the scalar variance of the random set.

Example 8 Let us now take again the random set $\Gamma = [T_n, T_x]$ from Examples 2 and 7, but let us now interpret it as a disjunctive random set providing incomplete information about the average daily temperature, T. Of course, the pairs of minimum and maximum temperatures may coincide during different days, even if their average temperatures do not. Conversely, the average temperatures during two different days may be the same, even if their respective pairs of min–max temperatures are not. For instance, the pair of min–max temperatures on a cloudy day may be, for instance, 17 and 21, and the pair on a clear day may be 12 and 22. The midpoints and especially the ranges are quite different, but the average temperatures can be pretty close. Besides, the midpoints of the respective intervals can be sometimes very far from the average temperature. According to the above ideas, the computation of some square distances between each of those intervals and the expected random interval might not provide very accurate information about the actual variance of the daily average temperature, Var(T). In contrast, the proper information about this variance should be given by a set of numbers, in accordance with Eq. 2.7, one of them being the actual variance:

$$\text{Var}(\Gamma) = \{\text{Var}(X) \colon X \in S(\Gamma)\}.$$

This was the proposal provided by Kruse in [32].

In each of the above examples, we have illustrated both alternative procedures introduced in the literature to extend the notion of parameter respectively recalled in paragraphs (A) and (B). The procedure described in (A) suits the interpretation of the random set as a random "object" as we have shown in Example 7, while the one described in (B) is the most appropriate procedure when the random set is regarded as the ill-observation of a real-valued random variable, according to Example 8.

The next example, taken from [8], shows that the scalar variance provides neither a lower bound nor an upper bound of Kruse's variance, in general.

Example 9 (a) The population $\Omega = \{\omega_1, \ldots, \omega_4\}$ comprises four objects, whose actual weights are $x_1^* = 10.2$, $x_2^* = 10.0$, $x_3^* = 10.4$, $x_4^* = 9.7$. We measure the weights with a digital device that rounds the measure to the nearest integer, and displays the value '10' in all of these cases. Therefore, we get the same interval of values for every object, $\gamma_i = [9.5, 10.5]$, $\forall i = 1, \ldots, 4$. The true variance of the four measurements is 0.067. But, we only know the information provided by the four intervals, so all we can say about the variance is that it is bounded by the values 0 and 0.25. (Kruse's variance returns this range of values.) On the contrary, the scalar variance (Körner, Lubiano and Feng definitions) returns the misleading value 0.

(b) Let us modify a bit the last situation. Let us suppose that, instead of four objects, we had only selected the first one (suppose that we had a population of size $N = 1$). Then, Kruse's variance would returns the singleton $\{0\}$. (We know with certainty that the variance of a single measurement is 0.) So, let us observe that Kruse's variance cannot be written as a function of the empirical distribution of the random set. In both cases [cases (a) and (b) in this example], the probability of the interval [9.5, 10.5] is 1, but Kruse's variance returns different ranges of values. On the contrary, Körner's and Feng's variances can be written as functions of the probability distribution induced by the random set. Hence, Körner's and Feng's variances would be useless within this context.

(c) Let us suppose that the four objects $\omega_1, \ldots, \omega_4$ have the same weight: $x_1^* = x_2^* = x_3^* = x_4^* = 9.8$, but, for some reason, the weight of the fourth object was measured with imprecision, and we only know that it is between the values 9.5 and 10.5. Our knowledge about the four measurements is given by $\gamma_1 = \{9.8\}$, $\gamma_2 = \{9.8\}$, $\gamma_3 = \{9.8\}$ and $\gamma_4 = [9.5, 10.5]$. The true variance in the sample is 0 and Kruse's variance produces the interval [0, 0.092]. According to the above incomplete information, Kruse's interval represents all our knowledge about the true variance. On the other hand, the scalar variance (Körner and Feng definitions) assigns a strictly positive value to it. We conclude that the scalar variance is neither an upper bound of the actual value of the variance of the underlying sample [see case (a) in this example], nor a lower bound [see case (c)].

As pointed out by Couso and Dubois [8], when the random set represents the imprecise observation of a standard random variable, the description of the changes of the observed sets via a scalar variance is not enough to inform about the variability of the underlying phenomenon, as we have shown in the last example. Nevertheless, the scalar variance of the random set may sometimes partially account for the variance of the underlying precise random variable. For instance, if a random set has disjoint outcomes the scalar variance is enough to reveal the non-deterministic nature of the underlying process (even if only partially). However, this is not always the case. It may only point out the variability *of the imprecision* of the observed outcomes, as it happens in Example 9 (even if the underlying quantity is not random) or when the outcomes are nested. On the other hand, as the above example shows, a zero scalar variance is not enough to conclude whether the observed phenomenon is random or not. Example 9a leads to a zero observable variance, because the variability of the weight is drowned in the imprecision of the observation.

The set of feasible values of the variance of the ill-known variable is not determined, in general, by the probability distribution induced by the random set as we have illustrated in the above example [part (b)]. This also happens with other parameters such as the entropy. Next we will try to illustrate what is the intuition behind this fact.

Example 10 Suppose that we have an urn with 3 balls, numbered from 1 to 3, which are coloured either red or white. The first ball is known to be red, the second one is white, and we do not know whether the third ball is red or white. We take a ball at random from the urn. We will receive \$1 if it is red, otherwise, we will not receive any reward. Let the random variable $X_0: \{1, 2, 3\} \to \{0, 1\}$ denote the reward obtained in this game, where the Laplace (discrete uniform) distribution is considered over the initial space. Our imprecise information about X_0 can be described by means of the random set $\Gamma: \{1, 2, 3\} \to \wp(\{0, 1\})$ where $\Gamma(1) = \{1\}$, $\Gamma(2) = \{0\}$ and $\Gamma(3) = \{0, 1\}$. Let us denote by means of a two-dimensional vector $(1 - p, p)$ any probability measure on $\{0, 1\}$ (where $1 - p$ and p respectively denote the mass assigned to 0 and 1). Then the probability envelope associated to Γ is the family

$$\mathcal{P}(\Gamma) = \{(1/3, 2/3), (2/3, 1/3)\}.$$

In other words, the probability measure induced by X_0 is a Bernoulli distribution whose parameter is either $1/3$ or $2/3$. In any case, the variance associated to X_0 is $\mathrm{Var}(X_0) = \frac{1}{3} \cdot \frac{2}{3} = \frac{2}{9}$ and its entropy, in bits, is exactly $H(X_0) = \frac{1}{3} \log_2 3 + \frac{2}{3} \log_2 \frac{3}{2}$. Therefore, in this example, both the "set-valued" variance and entropy of the random set, $\mathrm{Var}(\Gamma)$ and $\mathrm{H}(\Gamma)$, are singletons.

Let us now suppose that the urn has 150 balls numbered from 1 to 150. Suppose that the first 50 balls (one third) are known to be red, the second 50 balls are white, but, for each of the remaining 50 balls, we do not know whether it is red or white. Now consider again the same game as before: we take a ball at random from the urn and we receive \$1 if it is red, and nothing if it is white. Let us denote by $Y: \{1, \ldots, 150\} \to \{0, 1\}$ the random reward. Our imprecise information about Y can be represented by means of the random set $\Gamma': \{1, \ldots, 150\} \to \wp(\{0, 1\})$ defined as follows:

$$\Gamma'(i) = \begin{cases} \{1\} & \text{for } i = 1, \ldots, 50 \\ \{0\} & \text{for } i = 51, \ldots, 100 \\ \{0, 1\} & \text{for } i = 101, \ldots, 150. \end{cases}$$

The probability envelope of Γ', that represents the available information about the probability distribution of Y, is the family:

$$\mathcal{P}(\Gamma') = \left\{ \left(\frac{50+k}{150}, \frac{100-k}{150} \right) : k = 0, \ldots, 50 \right\}.$$

This means that all we know is that Y induces a Bernoulli distribution with parameter $p = \frac{100-k}{150}$, where k is an unknown integer belonging to $\{0, \ldots, 50\}$. Thus, according to the behavior of the function $f(p) = (1-p)p$ in the interval $\left[\frac{50}{150}, \frac{100}{150} \right]$ the lower and upper bounds for the variance of Y are respectively $\frac{1}{3} \frac{2}{3} = \frac{2}{9}$ and $\frac{1}{2} \frac{1}{2} = \frac{1}{4}$. Analogously, we can easily check that the entropy of Y, in bits, is an ill-known value belonging to the interval of values $[\frac{1}{3} \log_2 3 + \frac{2}{3} \log_2 \frac{3}{2}, 1]$.

Both random sets Γ and Γ' induce the same probability distribution on $\wp\{0, 1\}$. Both of them take each of the "values" $\{0\}$, $\{0, 1\}$ and $\{1\}$ with probability equal to $\frac{1}{3}$. However, while the first random set completely determines the values of the variance and the entropy of X_0, the second one just provides imprecise information about the variance and the entropy of Y. As we illustrate in this example, when the final space is finite, the "degree of atomicity" of the initial space impacts the width of the probability envelope, and therefore of the set-valued parameters associated to the random set. The "more atomic" the initial space, the more precise the information about them.

2.4.3 Relating Aumann Expectation and Choquet Integral

In 1953, Choquet [5] extended the notion of expectation (Lebesgue integral) to the case where the set function (measure) is not σ-additive but just monotone. Let $\mu \colon \wp(U) \to [0, 1]$ be a normalized ($\mu(U) = 1$) set-function satisfying the following restrictions:

- $\mu(\emptyset) = 0$.
- $A \subseteq B$ implies $\mu(A) \leq \mu(B)$ (monotony).

Let $g \colon U \to \mathbb{R}$ denote an $\mathcal{A}' - \beta_{\mathbb{R}}$ measurable mapping. We define the *Choquet integral* of g with respect to μ as:

$$(C) \int g \, d\mu = \int G_{\mu,g}(x) \, dx, \tag{2.8}$$

where $G_{\mu,g}(x) = \mu(\{u \colon g(u) \geq x\}) 1_{[0,\infty)} + [\mu(\{u \in U \colon g(u) \geq x\}) - 1] 1_{(-\infty,0)}$.

In particular, when μ is a probability measure the above integral can be expressed as follows:

$$(C) \int g \, d\mu = \int_0^\infty [1 - F_g(x)] \, dx - \int_{-\infty}^0 F_g(x) \, dx,$$

where F_g denotes the cumulative distribution induced by $g : U \to \mathbb{R}$. The above formula represents the Lebesgue integral (expectation) of g, $E_\mu(g)$. On the other hand, if μ is a 2-alternating capacity [inequality (2.5) with $k = 2$], the Choquet integral of g with respect to μ coincides with the supremum of the set of Lebesgue integrals of g with respect to those probability measures dominated by μ, i.e.:

$$(C) \int g \, d\mu = \sup\{E_Q(g) : Q \leq \mu\}.$$

Analogously, when μ is 2-monotone [inequality (2.6) with $k = 2$], the Choquet integral of g is the infimum of the expectations of g with respect to those probabilities that dominate μ. We can consider, in particular, the Choquet integrals of g with respect to the upper and lower probabilities induced by a random set, P^* and P_*. According to the above statement, the following equalities hold:

$$(C) \int g \, dP^* = \sup_{Q \in \mathcal{M}(P^*)} E_Q(g) \quad \text{and} \quad (C) \int g \, dP_* = \inf_{Q \in \mathcal{M}(P^*)} E_Q(g).$$

Furthermore, according to Miranda et al. [41], if g is bounded and $P^*(A) = \max \mathcal{P}(\Gamma)(A)$, $\forall A \in \mathcal{A}'$, it can be checked that:

$$(C) \int g \, dP^* = \max_{Q \in \mathcal{P}(\Gamma)} E_Q(g) \quad \text{and} \quad (C) \int g \, dP_* = \min_{Q \in \mathcal{P}(\Gamma)} E_Q(g).$$

Thus, if $g : \mathbb{R} \to \mathbb{R}$ is the identity function ($g(x) = x$, $\forall x \in \mathbb{R}$) and the image of Γ is bounded, we can deduce that those Choquet integrals do respectively coincide with the maximum and the minimum of Aumman's integral:

$$(C) \int g \, dP^* = \max E(\Gamma) \quad \text{and} \quad (C) \int g \, dP_* = \min E(\Gamma),$$

or equivalently that:

$$\max E(\Gamma) = \int_0^\infty P^*([x, \infty)) dx \; - \int_{-\infty}^0 P_*((-\infty, x)) \, dx$$

and

$$\min E(\Gamma) = \int_0^\infty P_*([x, \infty)) dx \; - \int_{-\infty}^0 P^*((-\infty, x)) \, dx.$$

2.5 Conditioning

Given a random set $\Gamma: \Omega \rightarrow \wp(U)$ and an event $A \in \mathcal{A}'$, the proper method for conditioning the random set on A depends on the adopted scenario.

2.5.1 Conditioning a Conjunctive Random Set

The problem of random set conditioning comes down to restricting the set-valued realisations $\Gamma(\omega)$ so as to account for the information that the set-valued outcome lies inside A. Then the obtained conditional mass is defined by means of the standard Bayes rule in the form of its weight distribution $p_\Gamma(\cdot|A)$ such that[5]:

$$p_\Gamma(C|A) = \begin{cases} \dfrac{p_\Gamma(C)}{\sum_{B \subseteq A} p_\Gamma(B)} & \text{if } C \subseteq A; \\ 0 & \text{otherwise,} \end{cases} \qquad (2.9)$$

where $p_\Gamma(B)$ denotes the probability that Γ takes the "value" B, i.e.:

$$p_\Gamma(B) = P(\{\omega \in \Omega : \Gamma(\omega) = B\}), \quad \forall B \in \text{Im}(\Gamma).$$

Example 11 Let S be a set of spoken languages including $c = $ Chinese, $e = $ English, $f = $ French and $s = $ Spanish. Let us consider the sets $C = \{e, f, s\}$ and $A = S \setminus \{c\}$. Then $p_\Gamma(C|A)$ denotes the proportion of people that speak English, French and Spanish (and nothing else), among those who cannot speak Chinese.

Equation (2.9) can be generalized by considering a pair of measurable families of subsets of U, $\mathcal{S} \in \sigma(\mathcal{C})$ and $\mathcal{A} \in \sigma(\mathcal{C})$ and writing

$$P_\Gamma(\mathcal{S}|\mathcal{A}) = \frac{P_\Gamma(\mathcal{S} \cap \mathcal{A})}{P_\Gamma(\mathcal{A})} = \frac{\sum_{B \in \mathcal{S} \cap \mathcal{A}} p_\Gamma(B)}{\sum_{A \in \mathcal{A}} p_\Gamma(A)}. \qquad (2.10)$$

Equation (2.10) is nothing else but the classical rule of conditioning, where the images of the random set Γ, which are subsets of U, are seen as "elements" of the actual "universe", and the families of subsets of U are the actual "events". In fact, in Eq. (2.9), we should write $p_\Gamma(C|\mathcal{A})$ where $\mathcal{A} = \{B : B \subseteq A\}$.

Example 12 Let U be again the set of spoken languages considered in Example 11 and let us consider the families of sets $\mathcal{S} = \{C : C \ni e\}$ and $\mathcal{A} = \{A : A \not\ni c\}$. Then, $p_\Gamma(\mathcal{S}|\mathcal{A}) = \frac{\sum_{C \in \mathcal{S} \cap \mathcal{A}} p_\Gamma(C)}{\sum_{A \in \mathcal{A}} p_\Gamma(A)}$ represents the proportion of people that can speak (at least) English, among those who do not speak Chinese.

---

[5] For the sake of simplicity, we will assume that the family of possible images of Γ, $\text{Im}(\Gamma)$, is finite. According to Shafer's Theory of Evidence [47], they will be referred to as "focal sets". This formulation can be easily extended to a more general context.

2.5.2 Conditioning an Ill-Known Random Variable: Prediction

Suppose the disjunctive random set Γ relies on a population Ω, and it induces the family of probabilities $\mathcal{P}(\Gamma)$ on $\mathcal{A}'$, one of which is the proper frequentist distribution of the underlying random variable X_0. Let $\mathcal{M}(P^*)$ denote its credal set. For the sake of simplicity, in this subsection the universe U will be assumed to be finite. Suppose we study a case ω_0 for which all we know is that $X_0(\omega_0) \in A$, and the problem is to predict the value of $X_0(\omega_0)$. Each probability $p_\Gamma(E)$ should be recomputed by restricting Ω to the subset $\Omega_A = \{\omega : X_0(\omega) \in A\}$ of population Ω relevant to case ω_0. However, because for each $\omega \in \Omega$, $X_0(\omega)$ is only known to lie in $\Gamma(\omega)$, the set Ω_A is itself ill-known.

Suppose we know for each focal set E, the proportion $\alpha_A(E)$ of the population (for which all we know is $X_0(\omega) \in E$) that lies inside Ω_A. Namely:

- If A and E are disjoint sets, $\alpha_A(E)$ will be equal to zero.
- Moreover, if $E \subseteq A$, then $\alpha_A(E) = 1$ necessarily.
- In the rest of the cases ($E \cap A \neq \emptyset$ and $E \cap A^c \neq \emptyset$), $\alpha_A(E)$ may take any value between 0 and 1.

We can obtain α_A if, for each non-empty subset E, we know the probability distribution p_E such that $p_E(u) = P(X_0(\omega) = u \mid \Gamma(\omega) = E)$ sharing the mass $p_\Gamma(E)$. It is clear that we can define $\alpha_{\{u\}}(E) = p_E(u)$, so that $\sum_{u \in E} \alpha_{\{u\}}(E) = 1$, $\forall E \in \text{Im}(\Gamma)$ and:

$$\alpha_A(E) = \sum_{u:\, u \in A} \alpha_{\{u\}}(E) = P_E(A), \quad \forall A \subseteq S, \forall E \in \text{Im}(\Gamma).$$

It defines a set $\mathbb{A} = \{\alpha_A(E) : A \subseteq U, E \in \text{Im}(\Gamma)\}$ of coefficients taking values on the unit interval $[0, 1]$. Each set $\mathbb{A}$ determines a probability measure, $Q \in \mathcal{M}(P^*)$ such that:

$$q(u) = \sum_{E \in \text{Im}(\Gamma)} \alpha_{\{u\}}(E) \cdot p_\Gamma(E), \quad \forall u \in U, \tag{2.11}$$

where $q : U \to [0, 1]$ denotes the weight distribution associated to Q.

If the set of coefficients $\{\alpha_A(E), E \in \text{Im}(\Gamma)\}$ is fixed, it is possible to condition p_Γ by restricting it to subsets of A as follows [14]:

$$p^{\alpha_A}(F|A) = \begin{cases} \dfrac{\sum_{F = E \cap A} \alpha_A(E) p_\Gamma(E)}{\sum_{E \cap A \neq \emptyset} \alpha_A(E) p_\Gamma(E)} & \text{if } F \subseteq A; \\ 0 & \text{otherwise.} \end{cases} \tag{2.12}$$

The idea is that the probability assigned to a subset F of A, when restricting to the population Ω_A is obtained by assigning to it all fractions $\alpha_A(E) p_\Gamma(E)$ of the population in Ω for which all we know is that $X_0(\omega) \in E$ and that we know to actually lie in $F = A \cap E$. It gives birth to ∞-alternating and monotone set functions

$\overline{P}^{\alpha_A}(B|A) = \sum_{F \cap B \neq \emptyset} p^{\alpha_A}(F|A)$ and $\underline{P}^{\alpha_A}(B|A) = \sum_{F \subseteq B} p^{\alpha_A}(F|A)$ over A. Varying the vector $(\alpha_A(E)_{E \subseteq S})$ leads to conjugate upper and lower set-functions as follows [14]:

$$\overline{P}(B|A) = \sup_{\alpha_A} \overline{P}^{\alpha_A}(B|A); \quad \underline{P}(B|A) = \inf_{\alpha_A} \underline{P}^{\alpha_A}(B|A). \quad (2.13)$$

These bounds are attained by the following choices of α_A vectors, where $B \subseteq A$:

- Upper bound $\overline{P}(B|A)$: $\alpha_A(E) = 1$ if $E \cap B \neq \emptyset$ or $E \subseteq A$, 0 otherwise.
- Lower bound $\underline{P}(B|A)$: $\alpha_A(E) = 1$ if $E \cap A \subseteq B$ or $E \subseteq A$, 0 otherwise.

In fact, it has been proved that the same bounds can be obtained by applying Bayesian conditioning to all probabilities in $Q \in \mathcal{M}(P^*)$ with $Q(A) > 0$. They are upper and lower conditional probabilities that take an attractive closed form [14, 22]:

$$\overline{P}(B|A) = \sup\{Q(B|A): Q \in \mathcal{M}(P^*)\} = \frac{P^*(B \cap A)}{P^*(B \cap A) + P_*(B^c \cap A)}, \quad (2.14)$$

$$\underline{P}(B|A) = \inf\{Q(B|A): Q \in \mathcal{M}(P^*)\} = \frac{P^*(B \cap A)}{P_*(B \cap A) + P^*(B^c \cap A)}, \quad (2.15)$$

where $\underline{P}(B|A) = 1 - \overline{P}(B^c|A)$ and B^c is the complement of B.

This is not surprizing since each vector α_A corresponds to a subset of probability measures $Q \in \mathcal{M}(P^*)$ obtained from all sets $\mathbb{A}$ of coefficients containing $\{\alpha_A(E): E \in \text{Im}(\Gamma)\}$ and each $Q \in \mathcal{M}(P^*)$ is generated by some set $\mathbb{A}$: more precisely $Q(B) = \sum_{E \subseteq U} p_\Gamma(E) \alpha_B(E)$ due to Eqs. (2.3), (2.4) and (2.11). Noticeably, $\overline{P}(B|A)$ and $\underline{P}(B|A)$ are still ∞-alternating and monotone, respectively, as proved in [29, 46], so that Eqs. (2.12) and (2.13) justify this form of conditioning (familiar in imprecise probability theory [49]) in the setting of belief functions.

The probability interval $[\underline{P}(B|A), \overline{P}(B|A)]$ represents what is known about the (frequentist) probability of B, restricting the probability space $(\Omega, \mathcal{A}, P)$, to the ill-known population of individuals $\omega \in \Omega$ for which $X_0(\omega) \in A$. Then, the interval $[\underline{P}(B|A), \overline{P}(B|A)]$ is what we can predict about the probability that $X_0(\omega_0)$ lies in B if we know that $\omega_0 \in A$.

The following example is a variant of Example 6 in [10].

Example 13 Suppose that we have three urns. The first one has 3 balls: one white, one red and one unpainted ($\Omega_1 = \{\omega_w^1, \omega_r^1, \omega^1\}$). The second urn has two balls: one red and one white ($\Omega_2 = \{\omega_w^2, \omega_r^2\}$). The third urn has two unpainted balls $\Omega_3 = \{\omega^3, \omega'^3\}$. We randomly select one ball ω_1 from the first urn. If it is colored, then we randomly select a second ball ω_2 from the second urn. Else if it is unpainted, we select the second ball ω_2 from the third urn. So the sample space is $\Omega = (\{\omega_w^1, \omega_r^1\} \times \Omega_2) \cup (\{\omega^1\} \times \Omega_3)$. Once the two balls have been selected, they will be both painted red or white according to an unknown procedure. The information about the final color of the pair of randomly selected balls can be represented by means of a random set Γ taking values on the product space $U \times U = \{r, w\} \times \{r, w\}$, denoting our

Table 2.1 Probability distribution of the random set Γ

γ	$P(\Gamma = \gamma)$
$\{(w, w)\}$	1/6
$\{(w, r)\}$	1/6
$\{(r, w)\}$	1/6
$\{(r, r)\}$	1/6
$\{(r, r), (w, w)\}$	1/3

Table 2.2 Mass distributions of the (measurable) selections of Γ

x	$P(\text{Selection } 1 = x)$	$P(\text{Selection } 2 = x)$
(w, w)	1/2	1/6
(w, r)	1/6	1/6
(r, w)	1/6	1/6
(r, r)	1/6	1/2

incomplete information about the final pair of colors. Γ is a set-valued functions with six possible outcomes. The probability of appearance of each of them is provided in Table 2.1.

The actual pair of colors of both (randomly selected) balls (ω_1, ω_2) can be represented by means of a random vector $X_0 = (X_1(\omega_1), X_2(\omega_2))$. All we know about it is that its values belong to the (set-valued) images of Γ, or, in other words, that the random vector X_0 belongs to the family of selections of Γ, $S(\Gamma)$. Actually, the random set Γ has only two different selections, whose probability distributions are provided Table 2.2.

Let us now consider, for instance the events $A = \{w\} \times U$ and $B = U \times \{w\}$. The conditional probability $p_X(B|A)$ denotes the probability that the color of second ball $(X_2(\omega_2))$ is also white, if we know that the color of the first one $(X_1(\omega_1))$ is white. Our knowledge about this value is determined by the pair of bounds

$$\overline{P}_\Gamma(B|A) = \sup_{X \in S(\Gamma)} p_X(B|A) = \frac{1/2}{1/2 + 1/6} = 3/4$$

and

$$\underline{P}_\Gamma(B|A) = \inf_{X \in S(\Gamma)} p_X(B|A) = \frac{1/6}{1/6 + 1/2} = 1/4.$$

As the setting is finite in the above example, $\overline{P}_\Gamma(B|A)$ and $\underline{P}_\Gamma(B|A)$ respectively coincide with the maximum and the minimum of the set of conditional probability values (in Eqs. (2.14) and (2.15)):

$$\{Q(B|A): Q \in \mathcal{M}(P^*), Q(A) > 0\} = \left\{ \frac{Q(B \cap A)}{Q(A)} : Q \in \mathcal{M}(P^*) \right\}.$$

Indeed, it has been proved in [7] that whenever $P^*(A)$ coincides with the maximum of $\mathcal{P}(\Gamma)(A) = \{P_X(A) : X \in S(\Gamma)\}$, for every $A \in \mathcal{A}'$, they also coincide with the maximum and the minimum of the set of values:

$$\{Q(B|A) : Q \in \mathcal{P}(\Gamma), Q(A) > 0\} = \left\{\frac{Q(B \cap A)}{Q(A)} : Q \in \mathcal{P}(\Gamma)\right\}.$$

2.5.3 Conditioning an Ill-Known Random Variable: Revision

Suppose we get on top of the random set $\Gamma : \Omega \to \wp(U)$ restricting the ill-known random variable X_0 on U, another piece of information improving our knowledge about the ill-known process, under the form of a subset $A \subseteq U$, such that actually $X_0(\omega) \in A, \forall \omega \in \Omega$. It means that we come to hear that the values taken by X_0 lie in A for sure. Then we perform an information fusion process that turns Γ into Γ_A such that $\Gamma_A(\omega) = \Gamma(\omega) \cap A, \forall \omega \in \Omega$ [16]. In other words, we have updated the set-valued mapping, improving the quality of the information. It yields yet another type of conditioning that systematically transfers probabilities $p_\Gamma(E)$ to $E \cap A$ when not empty. Two situations can be met.

- $\Gamma(\omega) \cap A \neq \emptyset, \forall \omega \in \Omega$, and therefore $P^*(A) = 1$, and $p_{\Gamma_A}(C) = \sum_{C = E \cap A} p_\Gamma(E)$. Then all upper probabilities $P^*(B)$ are turned into $P^*(A \cap B)$. It comes down to considering as impossible the values of $X_0(\omega)$ outside A.
- The set $(A^*)^c = (\Gamma^*(A))^c = \{\omega \in \Omega : \Gamma(\omega) \cap A = \emptyset\}$ is not null, and therefore $P^*(A) < 1$. It means that the initial information represented by means of Γ ($X_0(\omega) \in \Gamma(\omega), \forall \omega \in \Omega$) does not fully agree with the new information saying that $X_0(\omega) \in A$ for all $\omega \in A$. Mathematically speaking the mapping Γ_A is no longer in conformity with our initial setting as $\exists \omega \in \Omega, \Gamma_A(\omega) = \emptyset$. Then the solution proposed by Dempster is to condition the initial probability P on the set $A^* = \Gamma^*(A) = \{\omega : \Gamma_A(\omega) \neq \emptyset\}$, that is (Ω, P, Γ) is turned into (A^*, P_A, Γ_A), where $P_A = P(\cdot|A^*)$ denotes the (conditional) probability measure restricted to the collection of (measurable) events included in A^* as follows:

$$P_A(O) = P(O|A^*) = \frac{P(O)}{P(A^*)}, \forall O \subseteq A^*.$$

Note that this is what is usually done in standard probability theory. It comes down to considering the corresponding elements ω such that $\Gamma(\omega) \cap A = \emptyset$ as impossible, which may mean that the information we had on them was wrong.[6]

The resulting conditional probability on $\wp(U)$ in both cases is:

---

[6] But then strictly speaking, if we assume information A is correct, one may consider an alternative mapping Γ_A such that $\Gamma_A(\omega) = \Gamma(\omega) \cap A$ if not empty and $\Gamma_A(\omega) = A$ otherwise, that is, a genuine revision process; see [36].

$$pr_A(C) = \begin{cases} \frac{\sum_{C=E\cap A} pr(E)}{\sum_{E\cap A \neq \emptyset} pr(E)} & \text{if } C \subseteq A; \\ 0 & \text{otherwise,} \end{cases} \tag{2.16}$$

It leads to the alternative conditioning rule, called Dempster rule of conditioning [16]:

$$P^*(B|A) = \frac{P^*(A \cap B)}{P^*(A)} = \frac{P_\Gamma(\mathcal{F}_{A\cap B})}{P_\Gamma(\mathcal{F}_A)}, \tag{2.17}$$

and $P_*(B|A) = 1 - P^*(B^c|A)$. In fact, if $\forall \omega \in \Omega \, \Gamma_A(\omega) \neq \emptyset$, $P^*(B|A)$ is just the upper probability induced by (A^*, P_A, Γ_A). Note that it comes down to the previous conditioning rule (2.12) with $\alpha_A(E) = 1$ if $E \cap A \neq \emptyset$, and 0 otherwise (an optimistic assignment, justified by the claim that A contains the actual value of $X_0(\omega)$, $\forall \omega \in \Omega$). See Gilboa and Schmeidler [24] for a justification of this conditioning via the maximum likelihood principle (maximizing the probability of the observed event A).

The two conditioning rules are present in Dempster's pioneering paper [16], but only the latter is known as Dempster conditioning rule. Their meaning is radically different. We call rule (2.12) prediction conditioning, because it tries to predict the probability that the value $X_0(\omega_0)$ lies in B on a specific case ω_0 for which the population Ω is relevant, and for which all that is known is that $X_0(\omega_0) \in A$. We only use the available statistical knowledge to predict whether some singular case ω_0 will satisfy some property of interest, given what we know about this case. In contrast, the second conditioning rule (2.17) revises the available statistical knowledge, making it more precise, since we happen to hear that $X_0(\omega) \in A$ for all $\omega \in \Omega$.

Example 14 Let the random interval $\Gamma = [L, U]$ denote our incomplete information about the heights of the individuals in a population Ω. The random variables $L: \Omega \to \mathbb{R}$ and $U: \Omega \to \mathbb{R}$ respectively represent, for each individual, $\omega \in \Omega$, a lower and an upper bound of the actual height of ω. According to this information, we can provide a pair of upper and lower bounds for the proportion of individuals whose heights belong to the interval $B = [150, 170]$:

$$P_*(B) = P(\{\omega \in \Omega : \Gamma(\omega) \subseteq [150, 170]\}) = P(\{\omega \in \Omega : L(\omega) \geq 150, U(\omega) \leq 170\}),$$

and

$$P^*(B) = P(\{\omega \in \Omega : \Gamma(\omega) \cap [150, 170]\} \neq \emptyset) = P(\{\omega \in \Omega : L(\omega) \leq 170, U(\omega) \geq 150\}).$$

Suppose that we can add the following piece of information: "Every individual in the population is, at least, 160 cm tall." In that case, and if our initial information was compatible with this new statement (i.e., if $\Gamma(\omega) \cap [160, \infty) \neq \emptyset$, $\forall \omega \in \Omega$), we can provide refined information about the proportion of people between 150 and 170 (in fact, between 160 and 170 cm tall). These new bounds correspond to the lower and upper probabilities of B with respect to the random set $\Gamma_A = \Gamma \cap A$

which is well-behaved according to our original setting. In that case the interval $[P_*(B|A), P^*(B|A)] \subseteq [P_*(B), P^*(B)]$ indicating the increase in information.

For instance, suppose the simple situation where the population contains three individuals and that our initial information about their respective heights is given by the set-valued mapping $\Gamma(\omega_1) = \Gamma(\omega_2) = [140, 180], \Gamma(\omega_3) = [140, 165]$. Assume a uniform probability. Since $A = [160, +\infty), \Gamma_A(\omega_1) = \Gamma_A(\omega_2) = [160, 180], \Gamma_A(\omega_3) = [160, 165]$. Since $B = [150, 170], [P_*(B), P^*(B)] = [0, 1]$, while $[P_*(B|A), P^*(B|A)] = [1/3, 1]$.

Consider now the set-valued mapping $\Gamma'(\omega_1)=[140, 165], \Gamma'(\omega_2) = [140, 180], \Gamma'(\omega_3) = [140, 155]$. Again $[P_*(B), P^*(B)] = [0, 1]$. But now $(\Gamma')^*(A) = \{\omega_1, \omega_2\}$, which may mean that the information we had on ω_3 was wrong, or yet that it is an outlier. Now, $\Gamma'_A(\omega_1) = [160, 165], \Gamma'_A(\omega_2) = [160, 180]$, and now $[P_*(B|A), P^*(B|A)] = [0.5, 1]$.

2.6 Independence

The proper notion of independence between two random sets also depends on the specific interpretation (conjunctive or disjunctive) considered.

2.6.1 Random Set Independence

Under the conjunctive approach, two random sets are assumed to be independent if they satisfy the classical notion of independence, when considered as point-valued mappings, whose "values" are subsets of the final space. Mathematically, two multi-valued mappings $\Gamma_1 : \Omega \to \wp(U_1)$ and $\Gamma_2 : \Omega \to \wp(U_2)$, $\mathcal{A}$-$\sigma(\mathcal{C}_1)$ and $\mathcal{A}$-$\sigma(\mathcal{C}_2)$ measurable, respectively, are said to be independent if they satisfy the equalities:

$$P(\Gamma_1^{-1}(\mathcal{B}_1), \Gamma_2^{-1}(\mathcal{B}_2)) = P(\Gamma_1^{-1}(\mathcal{B}_1)) \cdot P(\Gamma_2^{-1}(\mathcal{B}_2)), \ \forall \mathcal{B}_1 \in \sigma(\mathcal{C}_1), \mathcal{B}_2 \in \sigma(\mathcal{C}_2).$$
$$(2.18)$$

Example 15 Let us consider two random sets defined on a population of people. Random set Γ_1 represents the set of spoken languages (one of them being e = English), and it assigns, to each randomly selected person, the set of languages (s)he can speak. Random set Γ_2 assigns to each person the set of countries (s)he has been to during more than two months (we will respectively denote by s and u the USA and the UK). Let us take a person at random and consider the events "(S)he speaks (at least) English" and "(S)he has been to the USA or the UK for more than two months". Those events can be mathematically expressed as $\Gamma_1 \ni e$ and $\Gamma_2 \cap \{s, u\} \neq \emptyset$, respectively. If they are not independent, i.e., if the probability of the conjunction

$P(\Gamma_1 \ni e, \Gamma_2 \cap \{s, u\} \neq \emptyset)$ does not coincide with the product of the probabilities of both events, $P(\Gamma_1 \ni e)$ and $P(\Gamma_2 \cap \{s, u\} \neq \emptyset)$, then we claim that Γ_1 and Γ_2 are stochastically dependent. In fact, in such a case, Γ_1 and Γ_2 would not satisfy Eq. 2.18 for $\mathcal{B}_1 = \{C : C \ni e\}$ and $\mathcal{B}_2 = \{D : D \cap \{s, u\} \neq \emptyset\}$ and therefore the random sets Γ_1 and Γ_2 are not stochastically independent.

2.6.2 Independence Between Ill-Known Variables

Suppose that we observe a pair of random variables X and Y with imprecision, and our information about each realization $(X(\omega), Y(\omega))$ is represented by a pair of sets, $\Gamma_1(\omega) \subseteq U_1, \Gamma_2(\omega) \subseteq U_2$, containing the actual pair of values. Given an arbitrary pair of events $A \subseteq U_1$ and $B \subseteq U_2$, the probability values $P_X(A)$ and $P_Y(B)$ are known to belong to the respective intervals of values: $[P_{*\Gamma_1}(A), P^*_{\Gamma_1}(A)]$ and $[P_{*\Gamma_2}(B), P^*_{\Gamma_2}(B)]$, where Γ_1 and Γ_2 denote the multivalued mappings that represent our incomplete information about X and Y, respectively, and $P_{*\Gamma_1}, P^*_{\Gamma_1}, P_{*\Gamma_2}, P^*_{\Gamma_2}$ denote their respective lower and upper probabilities, in the sense of Dempster. If we know, in addition, that both random variables are independent, then we can represent our knowledge about the joint probability $P_{(X,Y)}(A \times B)$ by means of the set of values:

$$\{p \cdot q : p \in [P_{*\Gamma_1}(A), P^*_{\Gamma_1}(A)] \quad \text{and} \quad q \in [P_{*\Gamma_2}(B), P^*_{\Gamma_2}(B)]\}.$$

According to [11], the family of joint probability measures:

$$\mathcal{P} = \{P : P(A \times B) = p \cdot q \text{ s.t. } p \in [P_{*\Gamma_1}(A), P^*_{\Gamma_1}(A)], q \in [P_{*\Gamma_2}(B), P^*_{\Gamma_2}(B)], \forall A, B\}.$$

is included in the family of probabilities dominated by the joint upper probability induced by the random set $\Gamma = \Gamma_1 \times \Gamma_2$, if we assume Γ_1 and Γ_2 to be stochastically independent, i.e., the upper probability defined as follows: $\forall A \in \mathcal{A}'_1, B \in \mathcal{A}'_2$,

$$\begin{aligned}
P^*_{\Gamma_1 \times \Gamma_2}(A \times B) &= P(\{\omega \in \Omega : \Gamma_1(\omega) \cap A \neq \emptyset, \Gamma_2(\omega) \cap B \neq \emptyset\}) \\
&= P(\{\omega \in \Omega : \Gamma_1(\omega) \cap A \neq \emptyset\}) \cdot P(\{\omega \in \Omega : \Gamma_2(\omega) \cap B \neq \emptyset\}) \\
&= P^*_{\Gamma_1}(A) \cdot P^*_{\Gamma_2}(B).
\end{aligned}$$

and likewise:

$$\begin{aligned}
P_{*\Gamma_1 \times \Gamma_2}(A \times B) &= P(\{\omega \in \Omega : \Gamma_1(\omega) \subseteq A, \Gamma_2(\omega) \subseteq B\}) \\
&= P(\{\omega \in \Omega : \Gamma_1(\omega) \subseteq A\}) \cdot P(\{\omega \in \Omega : \Gamma_2(\omega) \subseteq B\}) \\
&= P_{*\Gamma_1}(A) \cdot P_{*\Gamma_2}(B).
\end{aligned}$$

Furthermore, according to [10] $\mathcal{P}$ is, in general, a proper subset of the set of probability measures bounded by $P^*_{\Gamma_1 \times \Gamma_2}$ and $P_{*\Gamma_1 \times \Gamma_2}$. This fact does not influence the calculation of the lower and upper bounds for $P_{(X,Y)}(A \times B)$ in most common situations, but it impacts further calculations about bounds of other parameters associated to the probability measure $P_{(X,Y)}$ such as the variance or the entropy of some functions of the random vector (X, Y), as it is checked in [13].

As illustrated above, random set independence is the usual probabilistic notion in the conjunctive setting. But it has also a meaningful interpretation within the disjunctive framework: in this case, the pair of set-valued mappings (Γ_1, Γ_2) indicates incomplete information about an ill-observed random vector (X, Y). Independence between Γ_1 and Γ_2 indicates independence between incomplete pieces of information pertaining to the attributes (the sources providing them are independent), something that has nothing to do, in general, with the stochastic relation between the actual attributes X and Y themselves. We will illustrate this issue with some examples.

Example 16 Let us suppose that the set-valued mapping Γ_1 expresses vacuous information about an attribute X. In such a case, Γ_1 will be the constant set-valued mapping that assigns the whole universe, to any element ω in the initial space, and therefore, it is independent from any other set-valued mapping. But this fact does not mean at all that the attribute X is independent from every other attribute Y. Let us consider for instance the users of a crowded airport, and let X and Y respectively denote their height, in cm., and their nationality. Let us suppose that, on the one hand, we have precise information about their nationality, but on the other hand, we have no information at all about their height. We can represent our information about their height by means of the constant set-valued mapping Γ_1 that assigns the interval $[90, 250]$ to any passenger. On the other hand, our information about the nationality can be represented by means of a function whose images are singletons $(\Gamma_2(\omega) = \{Y(\omega)\}, \forall \omega \in \Omega)$. Both set-valued mappings satisfy the condition of random set independence but this does not mean that height and nationality are two independent attributes in the considered population.

Remark 3 Note that we may consider the extreme case where the two variables are the same $(X = Y)$ and there are two independent imprecise measurement processes, determined by two random sets $\Gamma_1 : \Omega \rightarrow \wp(U_1)$ and $\Gamma_2 : \Omega \rightarrow \wp(U_2)$, where $U_1 = U_2 = U$. Considering again the situation described in Example 14, for instance, suppose that $X = Y : \Omega \rightarrow \mathbb{R}$ represents the true height of the individuals in the population Ω. Then, the consistency between Γ_1 and Γ_2 requires that $\Gamma_1(\omega) \cap \Gamma_2(\omega) \neq \emptyset$, which is the expected situation. If this is not the case, some additional assumption on top of source independence must be accepted: if one admits that the random variable X under study should be a selection of $\Gamma = \Gamma_1 \cap \Gamma_2$, one is led to restrict our statistical setting to (Ω', P', Γ) where $\Omega' = \{\omega \in \Omega : \Gamma_1(\omega) \cap \Gamma_2(\omega) \neq \emptyset\}$, $P'(C) = P(C|\Omega')$. It means dropping some elements in the population Ω, considered as outliers, so as to justify a final conditioning on $\{\omega \in \Omega : \Gamma_1(\omega) \cap \Gamma_2(\omega) \neq \emptyset\}$. This overall procedure corresponds to Dempster rule of combination [16], and it has been used as a basic building block of Shafer's evidence theory [47]. Dempster rule of combination generalizes Dempster rule of conditioning to the case where the information provided about the individuals by the second source is not the same for all of them (namely, Dempster rule of conditioning results from assuming Γ_2 is a constant set-valued mapping, i.e., $\Gamma_2(\omega) = A, \forall \omega \in \Omega$).

The next example shows the converse situation: now, the ill-known information about a pair of independent random variables is represented by means of a pair of stochastically dependent random sets.

Example 17 Suppose that we have a light sensor that displays numbers between 0 and 255. We take 10 measurements per second. When the brightness is higher than a threshold (255), the sensor displays the value 255 during 3/10 s, regardless of the actual brightness value. Below we provide data for six measurements:

actual values	215	150	200	300	210	280
displayed quantities	215	150	200	255	255	255
set-valued information	{215}	{150}	{200}	[255, ∞)	[0, ∞)	[0, ∞).

The actual values of brightness represent a realization of a simple random sample of size $n = 6$, i.e., of a vector of six independent identically distributed random variables. Notwithstanding, our incomplete information about them does not satisfy the condition of random set independence. In fact, we have:

$$P(\Gamma_i \supseteq [255, \infty) | \Gamma_{i-1} \supseteq [255, \infty), \Gamma_{i-2} \not\supseteq [255, \infty)) = 1, \quad \forall i \geq 3.$$

In summary, when several random set-valued mappings indicate incomplete information about a sequence of random (point-valued) measurements of an attribute, independence between them would indicate independence between the sources of information, that we should distinguish from the actual independence between the actual outcomes. The difference between the idea of independence of several random variables and independence of the random sets used to represent incomplete information about them impacts the studies of Statistical Inference with imprecise data: according to the above examples, a sequence of random sets indicating imprecise information about a sequence of i.i.d. random variables does not necessarily satisfy the property of random set independence between its components. In fact, the family of product probabilities, each of which is dominated by an upper probability is strictly contained [10] in the family of probabilities dominated by the product of these upper probabilities (plausibilities, when the universe is finite). In other words, considering a sequence of i.i.d. random sets would not be appropriate when we aim to represent incomplete information about a sequence of i.i.d. random variables.

2.7 Exercises

1. Random sets can be given different interpretations as:

   • random objects (conjunctive ontic sets)
   • imprecisely known random variables (disjunctive epistemic sets)

   Choose the most appropriate of the above interpretations in the following cases:

   (a) The population is a set of people. The random set is a mapping assigning, to each person, the set of European countries (s)he has visited during the last 10 years.
   (b) The population is, again, a set of people. We have imprecise information about the number of countries each person has visited during the last 10 years. The random set assigns, to each person, our information about this number (a pair of lower and upper bounds for it).
   (c) The population is a collection of pregnant women. The random set assigns, to each of them, the interval of diastolic-systolic blood pressure at 9 am, in mmHg. The final goal is to study the distribution of this two-dimensional random vector (The pair of diastolic-systolic blood pressures in the population).
   (d) The population is a set of children. A psychologist is asked about the degree of dyslexia of each of them, in a range from 0 to 100. She provides, for each child, a more or less precise interval of values, according to their performance in different tests.

2. Consider a set of 10 students enrolled in an international course, $\Omega = \{s_1, \ldots, s_{10}\}$, and the following collection of languages:

   $$L = \{\text{Chinese, English, French, German, Italian, Spanish}\}$$

   The following multi-valued mapping reflects our knowledge about the number of those languages that each of the students can speak.

   $$\Gamma: \{s_1, \ldots, s_{10}\} \to \wp(\{1, 2, 3, 4, 5, 6\})$$

   $$\Gamma(s_1) = \{4, 5, 6\}, \ \Gamma(s_2) = \{2, 3\}, \ \Gamma(s_3) = \{2\}, \Gamma(s_i) = \{2, \ldots, 6\}, \quad i = 4, \ldots, 10.$$

   We consider the Laplace distribution over the initial set, representing the random selection of a student of the course.

   (a) What do we know about the proportion of students that speak 3 or more different languages? Which kind of random sets introduced in this chapter are we referring to?
   (b) Calculate the bounds of the Aumann expectation of Γ. What is the relation between both numbers and the actual expectation of the "number of languages spoken" in the population?

(c) Calculate a scalar variance of the convex hull of Γ, $\mathrm{Conv}(\Gamma) : \{s_1, \ldots, s_n\} \rightarrow$ $\wp([1, 6])$, defined as $\mathrm{Conv}(\Gamma)(\omega) = [\min \Gamma(\omega), \max \Gamma(\omega)]$, $\forall \omega \in \Omega$. (For instance, you can calculate the arithmetic mean of the variances of the extremes of Γ.)

(d) Calculate the bounds for the actual variance of the "number of languages spoken" in the population, according to the available information. Are they related to the scalar variance calculated before, or to some of the variances of the extremes of Γ?

3. Let $\Gamma : \Omega \rightarrow \wp(\mathbb{R})$ denote a strongly measurable multi-valued mapping defined on the set of the days of a year, representing the daily interval from the minimum to the maximum temperature, $\Gamma(\omega) = [T_n(\omega), T_x(\omega)]$. Check that the probability distribution of Γ is univocally determined by the joint distribution induced by T_n and T_x.

4. Let us consider a multi-valued mapping, $\Gamma : \Omega \rightarrow \wp(U)$, were U is a finite set. Prove that Γ is $\mathcal{A}$-$\wp(U)$ strongly measurable if and only if $\Gamma^{-1}(\{B\}) = \{\omega \in \Omega : \Gamma(\omega) = B\}$ is $\mathcal{A}$ measurable for every $B \in \wp(U)$.

5. Consider a non-empty random set $\Gamma : \Omega \rightarrow \wp(U)$ as an imprecise observation of a random variable $X_0 : \Omega \rightarrow \wp(U)$. Check that the upper and lower probabilities associated to Γ, P^* and P_*, and the probability measure induced by any measurable selection, $X \in S(\Gamma)$, satisfy the following inequalities:

$$P_*(A) \leq P_X(A) \leq P^*(A), \quad \forall A \in \mathcal{A}'.$$

Now assume that the universe U is finite and check that, for every $A \in \wp(U)$, there exists at least a pair (maybe coincident) of measurable selections $X_A : \Omega \rightarrow U$ and $Y_A : \Omega \rightarrow U$ such that $P_{X_A}(A) = P_*(A)$ and $P_{Y_A}(A) = P^*(A)$.

6. Consider an arbitrary finite universe U. In Evidence Theory [47] a mapping $m : \wp(U) \rightarrow [0, 1]$ is said to be a *basic mass assigment* when it satisfies the following restrictions:

- $m(\emptyset) = 0$
- $\sum_{A \subseteq U} m(A) = 1$.

Furthermore, the *belief* and the *plausibility measure* associated to m are the respective set-functions Bel: $\wp(U) \rightarrow [0, 1]$ and Pl: $\wp(U) \rightarrow [0, 1]$ defined as follows:

- $\mathrm{Bel}(B) = \sum_{A \subseteq B} m(A), \quad \forall B \in \wp(U)$
- $\mathrm{Pl}(B) = \sum_{A \cap B \neq \emptyset} m(A), \quad \forall B \in \wp(U)$.

Shafer's Evidence Theory and the theory of random sets are very closely related from a formal point of view.

(a) Consider a measurable space $(\Omega, \mathcal{A})$, a finite universe U and a $\mathcal{A} - \wp(U)$ measurable multi-valued mapping $\Gamma : \Omega \rightarrow \wp(U)$ with non-empty images. Check that the lower and upper probabilities associated to Γ do respectively coincide with the belief and plausibility measures associated to some mass assignment. Determine such a mass assignment as a function of P_Γ, the probability measure induced by Γ on $\wp(\wp(U))$.

(b) Consider a pair of belief and plausibility measures. Prove that there exists at least one random set whose lower and upper probabilities do respectively coincide with them.

7. Consider the following list, representing the actual values of a variable X on a sample of size 10:

$$2.1 \quad 4.3 \quad 4.2 \quad 1.7 \quad 3.8 \quad 7.5 \quad 6.9 \quad 5.2 \quad 6.7 \quad 4.8$$

Consider the grouping intervals $[0, 3)$, $[3, 6)$, $[6, 9]$, and draw the corresponding histogram of frequencies. Now suppose that someone else has imprecise information about the above data, and, instead of the exact quantities, he has just observed the following tuple of intervals, each of them containing the true outcome of the random variable of X:

$$[1, 4] \quad [2, 5] \quad [3, 5] \quad [1, 2] \quad [3, 5] \quad [4, 8] \quad [6, 8] \quad [4, 7] \quad [6, 8] \quad [3, 5]$$

Consider the same grouping intervals as before, and plot, for each interval, two lines, respectively corresponding to the maximum and the minimum frequency of the interval, according to the imprecise information obtained by this person. Compare this new "imprecise histogram" with the first one.

8. In Examples 9 and 10, we illustrated the fact that Kruse's variance cannot be written, in general, as a function of the probability distribution induced by a random set, when it is considered as a random object. In fact, we showed two pairs of random sets with the same probability distribution, but with different set-valued variances.

Now, consider the set $\Omega = \{h, t\}$, that denotes the possible outcomes in a coin tossing game and define the random sets $\Gamma_1 : \Omega \rightarrow \wp([-10, 10])$ and $\Gamma_2 : \Omega \rightarrow \wp([-10, 10])$ as follows:

- $\Gamma_1(h) = \Gamma_1(t) = [-10, 10]$
- $\Gamma_2(h) = [0, 10]$, $\Gamma_2(t) = [-10, 0]$.

Determine the probability distribution induced by each of the above random sets, as well as the interval-valued variance (according to Kruse's formula). Check that both random sets induce the same set-valued variance, but different probability distributions. Do they induce the same probability envelope or not?

9. Consider the two situations described in Example 10. Determine the credal set in both cases, and the pairs of bounds for the variance when we range the set of probability measures included in the credal set. Compare these bounds with those calculated in both parts of the example.

10. Let us consider a probability space $(\Omega, \mathcal{A}, P)$, the usual Borel σ-field $\beta_{\mathbb{R}}$ on the real line, and let $\Gamma : \Omega \rightarrow \wp(\mathbb{R})$ be a $\mathcal{A} - \beta_{\mathbb{R}}$ closed random interval mapping taking a finite number of (non-empty) interval-valued images,

$[a_1, b_1], \ldots, [a_n, b_n] \subseteq \mathbb{R}$ with respective probabilities $p_1, \ldots, p_n$. (i.e. $P(\Gamma = [a_i, b_i]) = p_i$, $i = 1, \ldots, n$). Consider the following interval of numbers:

$$\oplus_{i=1}^{n} [[a_i, b_i] \ominus E(\Gamma)]^2 \odot p_i, \qquad (2.19)$$

where the symbols $\oplus, \odot, \ominus$ represent the usual operations in interval-arithmetic, and $E(\Gamma)$ denotes the Aumann integral of the random interval. Now suppose that Γ represents the available imprecise information about an otherwise random variable $X : \Omega \rightarrow \mathbb{R}$ and check that the above interval contains the set of feasible values for the variance of X,

$$\mathrm{Var}(\Gamma) = \{\mathrm{Var}(Y) : Y \in S(\Gamma)\}.$$

Check that the interval considered in Eq. 2.19 is, in general, a proper super-set of $\mathrm{Var}(\Gamma)$.

11. The random variables X_0 and Y_0 respectively represent the temperature (in °C) of an ill person taken at random in a hospital just before taking an antipyretic (X_0) and 3 h later (Y_0). The random set Γ_1 represents the information about X_0 using a very crude measure (it reports always the same interval [37, 39.5]). The random set Γ_2 represents the information about Y_0 provided by a thermometer with ±0.5 °C of precision.

    (a) Are X_0 and Y_0 stochastically independent?
    (b) Are Γ_1 and Γ_2 stochastically independent?

12. (Example taken from [10]) Suppose that we have two urns, each of them with 10 balls. The first urn has five red, two white and three unpainted balls. The second one has three red, three white and 4 unpainted balls. We select one ball from each urn in a stochastically independent way, and if any of the selected balls is not coloured, then it is painted white or red by a completely unknown procedure. There can be arbitrary correlation between the colours they are finally assigned. Define a pair of random sets, Γ_1 and Γ_2, on $\{1, \ldots, 10\} \times \{1, \ldots, 10\}$ (the set of pairs of numbers associated to the pair of balls randomly selected) describing our knowledge about the color of both balls. The images of each of the above random sets will be, therefore, subsets of the set of colors {red, white}, denoting the set of possible colors of the corresponding ball. Determine the credal set associated to the Cartesian product random set $\Gamma = \Gamma_1 \times \Gamma_2$. Are Γ_1 and Γ_2 stochastically independent?

13. (Example taken from [10]) Consider again the urns from Exercise 12, and assume, for instance, the following procedure is used to assign colors to unpainted balls:

    • If only one of the selected balls is coloured, we will roll a dice to choose the colour of the other one. If the number on the dice is "5", we will paint it with the same colour. Otherwise, we will choose the opposite.

- If both selected balls have no colour we will roll two coins, each one for each ball. Each side of the coin will determine the final color of each ball.

Check that the resulting mappings $X_1, X_2 : \{1, \ldots, 10\} \times \{1, \ldots, 10\} \to \{r, w\}$ are not stochastically independent.

14. (Example taken from [10]) Suppose we have two urns, each one with 10 balls, numbered from 1 to 10. The first five balls in each urn are red, and the second five are unpainted. We choose a number i from 1 to 10 at random, and then we take the ball number i from each of the urns. (There is stochastic dependence between both selections.) Once we have selected both balls, they will be painted red or white, following a procedure which is completely unknown for us. Define a pair of multi-valued mappings, Γ_1 and Γ_2 on $\{1, \ldots, 10\}$ assigning, to each number i, our (imprecise) knowledge about the respective colors of both balls labeled with number "i". Check that Γ_1 and Γ_2 are not stochastically independent.

15. (Example taken from [10]) Consider the random experiment described in Exercise 14. Suppose that, once we have selected both balls, we use the following procedure to paint them in case they are uncolored. We toss three coins, and check the number of heads:

- If the number is 3, we paint both balls with the colour red.
- If the number of heads is 2, we paint the first ball red, and the second one, white.
- If the number of heads is 1, we paint the first ball white, and the second one, red.
- Finally, if three tails are obtained, we paint both of them white.

Determine the probability distribution of the random vector that represents the colors of the pair of the selected balls. Check that such a probability distribution can be factorized as the product of both marginals. In other words, check that the colors of both balls can be modeled as a pair of stochastically independent random variables.

16. Let the random variable $X : \Omega \to \mathbb{R}$ denote the actual weight of the objects in a set Ω. Suppose that there is some device that takes one object per second at random from the set Ω and places it on a scale. Due to some fault in the scale, when the weight of an object gets over a threshold $k \in \mathbb{R}$, the scale keeps displaying the message error during the next 3 s. Let $(X_1, \ldots, X_n)$, representing the actual weights of n objects taken at random (with replacement) during n consecutive seconds. Let $(\Gamma_1, \ldots, \Gamma_n)$ denote our information about the weights of those n objects. Check that the random sets $\Gamma_1, \ldots, \Gamma_n$ are not stochastically independent. (The imprecise information about a simple random sample of variables is not, in general, modeled by means of a tuple of stochastically independent random sets.)

References

1. J. Aumann, Integral of set valued functions. J. Math. Anal. Appl. **12**, 1–12 (1965)
2. S. Bochner, Integration von Funktionen, deren Werte die Elemente eines Vectorraumes sind. Fundam. Math. **20**, 262–276 (1933)
3. C. Castaing, M. Valadier, *Convex Analysis and Measurable Multifunctions* (Springer, Berlin, 1977)
4. A. Castaldo, F. Maccheroni, M. Marinacci, Random correspondences as bundles of random variables. Sankhya Ind. J. Stat. **66**, 409–427 (2004)
5. G. Choquet, Theory of capacities. Annales de l'Institut Fourier. University of Grenoble **5**, 131–295 (1953)
6. G. de Cooman, D. Aeyels, Supremum preserving upper probabilities. Inf. Sci. **118**, 173–212 (1999)
7. I. Couso, in *Teoría de la probabilidad para datos imprecisos. Algunos aspectos*. Ph.D. thesis, University of Oviedo, 1999 (in Spanish)
8. I. Couso, D. Dubois, On the variability of the concept of variance for fuzzy random variables of variance. IEEE Trans. Fuzzy Syst. **17**, 1070–1080 (2009)
9. I. Couso, D. Dubois, Statistical reasoning with set-valued information: ontic versus epistemic views. Int. J. Approximate Reasoning (2014). doi:10.1016/j.ijar.2013.07.002
10. I. Couso, S. Moral, Independence concepts in evidence theory. Int. J. Approximate Reasoning **51**, 748–758 (2010)
11. I. Couso, S. Moral, P. Walley, A survey of concepts of independence for imprecise probabilities. Risk Decis Policy **5**, 165–181 (2000)
12. I. Couso, L. Sánchez, P. Gil, Imprecise distribution function associated to a random set. Inf. Sci. **159**, 109–123 (2004)
13. I. Couso, L. Sánchez, Higher order models for fuzzy random variables. Fuzzy Sets Syst. **159**, 237–258 (2008)
14. L.M. de Campos, M.T. Lamata, S. Moral, The concept of conditional fuzzy measure. Int. J. Intell. Syst. **5**, 237–246 (1990)
15. G. Debreu, Integration of correspondences, in *Proceedings of the Fifth Berkeley Symposium of Mathematical Statistics and Probability* (Berkeley, USA, 1965), pp. 351–372
16. A.P. Dempster, Upper and lower probabilities induced by a multi-valued mapping. Ann. Math. Stat. **38**, 325–339 (1967)
17. D. Denneberg, *Non-additive Measure and Integral* (Kluwer Academic Publishers, Dordretch, 1994)
18. D. Dubois, Possibility theory and statistical reasoning. Comput. Stat. Data Anal. **51**, 47–69 (2006)
19. D. Dubois, H. Prade, *Possibility Theory* (Plenum Press, New York, 1988)
20. D. Dubois, H. Prade, Formal representations of uncertainty, in *Decision-Making Process Concepts and Methods*, ed. by D. Bouyssou, D. Dubois, M. Pirlot, H. Prade (ISTE & Wiley, London, 2009)
21. D. Dubois, H. Prade, Gradualness, uncertainty and bipolarity: making sense of fuzzy sets. Fuzzy Sets Syst **192**, 3–24 (2012)
22. R. Fagin, J.Y. Halpern, A new approach to updating beliefs, in *Uncertainty in Artificial Intelligence*, ed. by P.P. Bonissone, M. Henrion, L.N. Kanal, J.F. Lemmer (1991)
23. Y. Feng, L. Hu, H. Shu, The variance and covariance of fuzzy random variables and their applications. Fuzzy Sets Syst. **120**, 487–497 (2001)
24. I. Gilboa, D. Schmeidler, Updating ambiguous beliefs. J. Econ. Theory **59**, 33–49 (1993)
25. I.R. Goodman, H.T. Nguyen, *Uncertainty Models for Knowledge-Based Systems* (Elsevier Science Publishers, Amsterdam, 1985)
26. C. Hess, Mesurabilité, convergence et approximation des multi-fonctions à valeurs dans un e.l.c.s., Technical Report 15, Séminaire d'Analyse Convexe, Université du Languedoc, Montpellier, France, 1985. Exp 9, 100 pp

27. W. Hildenbrand, *Core and Equilibria of a Large Economy* (Princeton University Press, Princeton, 1974)
28. C.J. Himmelberg, Measurable relations. Fund. Math. **87**, 53–72 (1975)
29. J.Y. Jaffray, Bayesian updating and belief functions. IEEE Trans. Syst. Man Cybern. **22**, 1144–1152 (1992)
30. D.G. Kendall, Foundations of a theory of random sets, in *Stochastic Geometry*, ed. by E.F. Harding, D.G. Kendall (Wiley, New York, 1974), pp. 322–376
31. R. Körner, On the variance of fuzzy random variables. Fuzzy Sets Syst. **92**, 83–93 (1997)
32. R. Kruse, On the variance of random sets. J. Math. Anal. Appl. **122**, 469–473 (1987)
33. R. Kruse, K.D. Meyer, *Statistics with Vague Data* (D. Reidel Publishing Company, Dordrecht, 1987)
34. M.A. Lubiano, *Variation measures for imprecise random elements*, Ph.D. thesis, University of Oviedo, 1999 (in Spanish)
35. N. Lusin, *Leçons sur les ensembles analytiques et leurs applications* (Chelsea, New York, 1930)
36. J. Ma, W. Liu, D. Dubois, H. Prade, Bridging Jeffrey's rule, AGM revision and Dempster conditioning in the theory of evidence. Int. J. Artif. Intell. Tools **20**(4): 691–720 (2011)
37. G. Matheron, *Random Sets and Integral Geometry* (Wiley, New York, 1975)
38. E. Miranda, I. Couso, P. Gil, Relationships between possibility measures and nested random sets. Inf. Sci. **10**, 1–15 (2002)
39. E. Miranda, I. Couso, P. Gil, Random sets as imprecise random variables. J. Math. Anal. Appl. **307**, 32–47 (2005)
40. E. Miranda, I. Couso, P. Gil, Random intervals as a model for imprecise information. Fuzzy Sets Syst. **154**, 386–412 (2005)
41. E. Miranda, I. Couso, P. Gil, Approximation of upper and lower probabilities by measurable selections. Inf. Sci. **180**, 1407–1417 (2010)
42. I. Molchanov, *Limit Theorems for Unions of Random Closed Sets*. Lecture Notes in Mathematics, vol. 1561 (Springer, Berlin, 1993)
43. I. Molchanov, *Theory of Random Sets* (Springer, London, 2005)
44. H.T. Nguyen, On random sets and belief functions. J. Math. Anal. Appl. **63**, 531–542 (1978)
45. P. Novikov, Sur les fonctions implicites mesurables B. Fund. Math. **17**, 8–25 (1931)
46. J. Paris, *The Uncertain Reasoner's Companion* (Cambridge University Press, Cambridge, 1994)
47. G. Shafer, *A Mathematical Theory of Evidence* (Princeton University Press, Princeton, 1976)
48. D. Stoyan, W.S. Kendall, J. Mecke, *Stochastic Geometry and its Applications* (Wiley, Chichester, 1995)
49. P. Walley, *Statistical Reasoning with Imprecise Probabilities* (Chapman and Hall, London, 1991)
50. R. Yager, L.-P. Liu (eds.), *Classic Works of the Dempster–Shafer Theory of Belief Functions*, vol. 219, Studies in Fuzziness and Soft Computing (Springer, Berlin, 2008)

Chapter 3
Random Fuzzy Sets as Ill-Perceived Random Variables

3.1 Introduction: Two Views of Fuzzy Random Variables

The concept of fuzzy random variable, that extends the classical definition of random variable, was introduced by Féron [37] in 1976.[1] Later on, several authors, and especially Kwakernaak [54], Puri and Ralescu [62], Kruse and Meyer [53], Diamond and Kloeden [26], proposed other variants. More recently Krätschmer [51] surveyed all of these definitions and proposed a unified formal approach. In all of these papers, a fuzzy random variable is defined as a function that assigns a fuzzy subset to each possible output of a random experiment. Just like in the particular case of random sets considered in Chap. 1, the different definitions in the literature disagree on the measurability conditions imposed to this mapping, and in the properties of the output space, but all of them intend to model situations that combine fuzziness and randomness.

Like for random sets, there is not a unique intuition behind the various definitions originally proposed in the literature. While Feron [37], Puri and Ralescu [62], Diamond and Kloeden [26] view a fuzzy random variable as the extension of a random set to a random membership function, Kwakernaak [54], Kruse and Meyer [53] consider that this membership function models the imprecise perception of an ill-known classical random variable. This divergence of views, inherited from the same one existing for random sets, again directly impacts the choice of suitable definitions for extensions of traditional characteristic parameters.

In this section, we shall briefly review these two interpretations. Each of them is in accordance with a specific view of fuzzy sets (the so-called "ontic" and "epistemic" interpretations already pointed out in the previous chapters for sets). In the next subsection, we shall clarify the difference between both views.

---

[1] For a presentation of Féron's works and a bibliography, see [2].

© The Author(s) 2014   47
I. Couso et al., *Random Sets and Random Fuzzy Sets as Ill-Perceived Random Variables*,
SpringerBriefs in Computational Intelligence, DOI: 10.1007/978-3-319-08611-8_3

3.1.1 The Different Semantics of Fuzzy Sets

In Chap. 1, we reviewed and exemplified the so-called conjunctive and disjunctive interpretations of sets. A conjunctive set is a collection of elements representing precise information about an objective entity (the set of languages that some person can speak, or the interval of temperatures, from the min to the max, on a specific date, or a region in a digital image, etc.), while a disjunctive set only represents incomplete information about a certain element (our imprecise knowledge about the birth date of a person, of about the number of languages (s)he can speak, etc.).

Regardless of the particular interpretation, the formal notion of "(crisp) set" is a unique and well-defined concept. Furthermore, any set can be univocally determined by means of its "characteristic function", i.e., the mapping defined on the whole universe assigning the value 1 to all the elements in the set, and the value 0, to the rest. Similarly, a fuzzy set is modeled by a mapping, called its "membership function", which is defined on the universe and takes values in a totally ordered set L, which is usually the unit interval. Thus, the mathematical definition of a fuzzy set is unique (like the mathematical notion of a crisp set), but it needs to be interpreted in practical problems, in order to be used meaningfully. We can extend the distinction between "conjunctive" and "disjunctive" sets to fuzzy sets, and speak about "ontic" and "epistemic" fuzzy sets, respectively. Those terms have been proposed by Dubois and Prade (see [36], for instance) in analogy with ontic versus epistemic actions in cognitive robotics [46]. According to the authors, ontic fuzzy sets represent objects originally construed as sets, but for which a fuzzy representation is more expressive due to gradual boundaries. Degrees of membership evaluate to what extent components participate to the global entity. In contrast, the "epistemic" interpretation regards fuzzy sets as possibility distributions representing incomplete information about elements of the universe. In the same way we did in Chap. 1, we shall illustrate the difference between both views with some examples. Afterwards, we shall focus on the second (epistemic) interpretation of fuzzy sets.

Example 1 Let us consider the same set of languages as in Chap. 1,

$$L = \{\text{Dutch, English, French, German, Italian, Russian, Spanish}\}.$$

We can associate a fuzzy subset of L to each person, reflecting precise information about her/his language skills. Thus, for each person, and each language, a membership value indicating a degree of proficiency of this person in speaking that language on a [0, 1]-scale can be assigned. Those degrees can be determined as functions of the CEFR levels,[2] for instance.

Example 2 A person has a container of apples and he is asked about their weight. He does not use any scale. He just takes an apple in his hand and chooses between "high", "medium" and "low", these terms forming the linguistic scale. Each of these

---

[2] CEFR stands for Common European Framework of Reference for Languages.

labels can be viewed as a suitable fuzzy subset of the weight scale. According to [22, 33], the membership of an arbitrary weight x to each of the above fuzzy sets (those associated to the respective labels "high", "medium", "low") could represent the probability of selecting the corresponding label, for an apple whose weight is x. Thus, the respective membership values could be objectively calculated as follows: each membership value $\mu_i(x)$ would represent the proportion of people that assign the label i ($i = H, M, S$) to the weight x. In such a case, the family of labels would be associated to a strong [63] fuzzy partition of the universe.

In Examples 1 and 2, we have considered the "ontic" interpretation of fuzzy sets. Now, let us exemplify the epistemic approach.

Example 3 A vehicle moves from one point to another. A GPS detects its position with some imprecision. In fact, it attaches confidence degrees to a family of nested circles. The information provided by the GPS about a specific position can be represented by means of a fuzzy set (understood as a possibility measure).

3.1.2 Different Interpretations of Fuzzy Random Variables

Each of the above interpretations of fuzzy sets leads to a different interpretation of fuzzy random variable. We shall call them the "random fuzzy sets" and the "ill-known random variable" approaches, respectively. We shall proceed to briefly review both approaches.

In [62], Puri and Ralescu consider that the observations of some random experiments do not consist of numerical outputs, but are represented by fuzzy sets representing for instance vague linguistic terms (a linguistic variable in the sense of Zadeh [79]). According to this idea, some authors consider that a fuzzy random variable is a measurable function, in the classical sense, between a certain σ-algebra of events in the original space and a σ-algebra defined over a class of fuzzy subsets of $\mathbb{R}$. In this context, the probability distribution induced by the fuzzy random variable can be used to summarize the probabilistic information that the variable provides. If the fuzzy random variable has a finite number of images forming a linguistic term set, probability values can be assigned to the different linguistic labels, thus forming a random fuzzy set. For example, the following model could be generated: the result is "high" with probability 0.5, "medium" with probability 0.25 and "low" with probability 0.25, where "high", "medium" and "low" are linguistic labels associated to fuzzy subsets of some numerical scale.

Within this framework, we can use the tools of general Probability Theory to define the probability distribution induced by a fuzzy random variable (FRV). We can also extend the concepts of expectation, variance, etc. by reproducing classical techniques. For instance, when the images of the FRV are convex fuzzy subsets of $\mathbb{R}$, we can use fuzzy arithmetic to derive a method of construction of the expectation: a limit-based construction analogous to Lebesgue integral definition (using Zadeh's

Extension Principle for the sums and products needed to define the expectation of "simple" FRV) should lead to a definition of expectation which is consistent with Puri and Ralescu's one [62]. This expectation is a fuzzy subset of the final space and it plays the role of the "average value" of the FRV. On the other hand, we can make a parallel construction of the variance: let us consider a particular metric defined over the class of fuzzy subsets of the final space. In this setting, we can define the variance of an FRV as the mean (classical expectation of a random variable) of the squares of the distances from the images of the FRV to the (fuzzy) expectation. The respective definitions of variance given by Feng [40], Körner [50] and Lubiano [56] fit this formulation. In this context the variance of an FRV is a (precise) number that quantifies the degree of dispersion of the membership functions obtained as images of the FRV.

Thus, from a purely formal point of view, this "classical" view of the probability model induced by a fuzzy random variable allows us to use the tools of Probability Theory to transfer different classical concepts and results from the Probability Theory to this new environment. If, in particular, the fuzzy random variable has a finite number of images, probability values can be assigned to different fuzzy labels, as we exemplified above. Then, we can assign a precise probability value to each (measurable) sub-class of fuzzy subsets.

Unfortunately, this kind of mathematical model is not useful in some problems where the images of the fuzzy random variable represent the imprecise observations of the outcomes of a random experiment [9, 64–69]. It does not reflect the available (imprecise) information about the "true" probability distribution that governs the random experiment. Our imprecision should be reflected in our knowledge about the probability of each event. These events are in fact crisp subsets of the final space, but our information about their probability of occurrence is imprecise. Hence we should look for an imprecise model that assigns an imprecise quantity (a crisp or a fuzzy subset of the unit interval) to each particular event (measurable crisp set of the final space). Furthermore, the extensions of the concepts of expectation and variance should reflect, under this interpretation, our imprecise knowledge about the true (crisp) values of the expectation and variance. Both of them should be defined as imprecise quantities and not as precise numbers.

With this idea in mind, Kruse and Meyer [53] choose a possibilistic interpretation of fuzzy sets. Each fuzzy set is viewed as modeling incomplete knowledge about an otherwise precise value. These authors then claim that the fuzzy random variable represents imprecise or vague knowledge about a *classical* random variable, X_0 : $\Omega \to \mathbb{R}$, they refer to as the "original random variable." Therefore, the membership degree of a point x to the fuzzy set $\tilde{X}(\omega)$ represents the possibility degree (interpreted as its plausibility) of the assertion "$X_0(\omega) = x$", i.e., the image of element ω coincides with x. This way, the authors get all the elements needed to define a possibility measure and a dual necessity measure over the set of all random variables. They do it via the so-called "acceptability degree" of each random variable, that takes values in the unity interval. Actually, the acceptability function can be regarded as a possibility distribution defined over the set of all random variables. The degree of acceptability of X represents the plausibility (possibility degree) that X is the "true" random variable that models the studied experiment. If the fuzzy random variable were a random set (its images are crisp subsets of $\mathbb{R}$) the acceptability function would assign the value 1 to random variables that stand as selections of the random set (as defined in Sect. 2.3) and the value 0 to the remaining ones. When, in particular, the fuzzy random variable is a classical random variable (all the images are singletons) the acceptability function would assign the value 1 to only one random variable, which is the true random variable that models the experiment. In this case, its observation is completely precise.

In this setting, we can induce, in a natural way, an acceptability function over the class of all possible probability distributions on the range of the ill-known random variable. It represents the available information about the true probability distribution of the random experiment under study. From a theoretical point of view, this model is equivalent to a second-order possibility distribution [16]: the information about the true probability is represented by a possibility measure defined over the class of all probability measures. As said in the previous subsection, a possibility measure can encode a set of (meta-)probability measures (it is the set of probability measures it dominates). Therefore, a second-order possibility measure is associated to a set of probability measures, each one of them is defined, in turn, over a set of probability measures. Thus, a second-order possibility will allow us to state assertions like "the probability that the true probability of the value 7 is 0.5 is at least 0.7", etc.

In Sect. 3.4, we shall study this model in detail, in order to present the best way to express our incomplete information about the probability measure induced by X_0, and about the most common parameters, such as the expectation or the variance.

3.2 Prerequisites

In this chapter, we shall focus on fuzzy random variables that represent incomplete knowledge about an otherwise precise random variable. Under this view, the values of the random variables are ill-known and represented by fuzzy sets that convey an epistemic flavour. In order to specify how to construct membership functions under this approach, we shall show the mathematical relation between families of nested confidence regions, possibility distributions and membership functions. And we try to provide a precise meaning to the idea of fuzzy (set-valued) probability.

3.2.1 Possibility and Necessity Measures

First, let us recall the notions of *possibility distribution* and *possibility measure* [31]:

Definition 1 Given an arbitrary referential set U, a possibility measure is a set-function $\Pi : \wp(U) \to [0, 1]$ satisfying the following constraints:

- There exists a (unique, by construction) mapping $\pi : \quad U \to [0, 1]$ such that $\Pi(A) = \sup_{x \in A} \pi(x), \forall A \subseteq U$.

- $\Pi(U) = \sup_{x \in U} \pi(x) = 1$.

The mapping π is called the possibility distribution associated to Π.

The conjugate of a possibility measure is called a necessity measure, denoted by N and such that $N(A) = 1 - \Pi(A^c)$. The degree $N(A)$ can be interpreted as a degree of certainty of A on the basis of the knowledge described by π. While $\Pi(A)$ is a degree of consistency of A with π (evaluating a form of overlap between them), $N(A)$ is a degree of entailment of A from π. As a consequence of these definitions $\Pi(A \cup B) = \max(\Pi(A), \Pi(B))$ and $N(A \cap B) = \min(N(A), N(B))$. The degree $\pi(u)$ should be understood as quantifying the plausibility of u as being the true value of some ill-known quantity X (or equivalently it quantifies the lack of surprise at hearing that $X = u$ [70]); $\Pi(A)$ also evaluates the lack of surprise caused by $X \in A$.

Let the reader notice that, according to the last definition, a possibility distribution π is, formally speaking, the membership function of a normal fuzzy set. According to this, any normal fuzzy set can univocally represent a possibility measure. Under the epistemic stance, the larger a disjunctive set the less informative. This view naturally extends to possibility distributions, noticing that π is less specific (that is, less informative, less committed) than π' whenever $\pi \geq \pi'$.

In the framework of imprecise probabilities, a possibility measure is a coherent upper probability, i.e., it is the supremum of a convex set of (finitely additive) probability measures on $\wp(U)$. Such a convex set is called the credal set associated to Π or, equivalently, the family of probability measures dominated by Π:

$$\mathcal{P}_\Pi = \{P : P(A) \leq \Pi(A), \forall A \subseteq U\}.$$

The following result, which is taken from [14], shows the relationship between a nested family of confidence regions and a possibility distribution (see also [27, 32]).

Theorem 1 *Consider a possibility distribution, π, associated to a possibility measure Π, and the induced convex set $\mathcal{P}_\Pi$, we have that $P \in \mathcal{P}_\Pi$ if and only if $1 - \alpha \leq P(\{x \in U : \pi(x) > \alpha\})$, for all $\alpha \in (0, 1]$.*

According to the last theorem, we can use a fuzzy set $\tilde{A}$ to represent incomplete probabilistic information about an element $x_0 \in U$. Such an incomplete information admits the following three equivalent representations:

- For every $x \in U$, the possibility degree that x_0 coincides with x is $\tilde{A}(x)$.
- For every $B \subseteq U$, the probability that x_0 belongs to B is less than or equal to $\Pi(B) = \sup_{x \in B} \tilde{A}(x)$.
- The probability that x_0 belongs to $\tilde{A}_{(\alpha)} = \{x \in U : \tilde{A}(x) > \alpha\}$ is greater than or equal to $1 - \alpha$, $\forall \alpha \in (0, 1]$.

Conversely, a family of nested sets with prescribed confidence levels is well-known to directly yield a fuzzy set. A *gradual set* (*graded* in the terminology of [45][3]) of U is a multi-valued mapping $\varphi : [0, 1] \to \wp(U)$ satisfying

---

[3] It is called a gradual set, without any nestedness requirement in [35].

$$\forall \alpha, \beta \in [0, 1], [\alpha \leq \beta \Rightarrow \varphi(\alpha) \supseteq \varphi(\beta)].$$

For an arbitrary gradual set φ, there exists a unique fuzzy set, $\pi_\varphi : U \to [0, 1]$ that satisfies:

$$\pi_\varphi(u) = \sup\{\alpha : u \in \varphi(\alpha)\}, \forall u \in U.$$

In many practical situations, such as the one considered in Example 3, we are provided with incomplete information about a certain point in the universe, which is represented by means of a family of nested sets, each one of them associated to a certain confidence degree. According to Theorem 1, we can use normal fuzzy sets to represent this kind of imprecise information. Let us illustrate the construction of a fuzzy set in a specific situation.

Example 4 Suppose that we take an apple from a container. We use a scale, that is "under control" 90 % of the time, where measurements are within a 10g error margin. In the remaining 10 % of the time, the scale is "out of control" and we can only guarantee an error less than 50g. If x_0 denotes the true weight of the apple, and d represents the displayed value, our information about x_0 can be described by the fuzzy number $\tilde{A}$:

$$\tilde{A}(x) = \begin{cases} 0 & \text{if } x \notin [d - 50, d + 50] \\ 0.1 & \text{if } x \in [d - 50, d - 10) \cup (d + 10, d + 50]. \\ 1 & \text{if } x \in [d - 10, d + 10]. \end{cases}$$

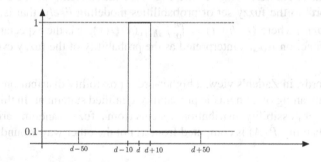

Throughout the rest of this chapter, we shall use π to denote a possibility distribution and the membership function of its associated fuzzy set. For any $\alpha \in [0, 1]$ we shall denote by π_α the (weak) α-cut of π, i.e. $\pi_\alpha = \{u : \pi(u) \geq \alpha\}$. We shall say that a possibility measure Π is *normal* when $\Pi(U) = 1$.[4] We shall denote by $\mathcal{F}(U)$ the class of all fuzzy subsets of U.

---

[4] This is completely unrelated to the idea of normal distribution in probability, that is, the use of Gaussian curves.

3.2.2 Fuzzy-Valued Probabilities of Events

A *possibilistic probability* (or a "fuzzy probability") [18], $\tilde{P} : \mathcal{A}' \to \mathcal{F}([0, 1])$, is a map taking each event $A \in \mathcal{A}'$ to a normal possibility distribution, $\tilde{P}(A)$ on $[0, 1]$. The idea of imprecisely describing probability degrees of events by fuzzy sets is not new and was proposed by Zadeh [81], who later considered them as coarse, qualitative, or "granular" probabilities [82] (often representing linguistic probability terms like *probable, unlikely*, etc.)

Here, we view the concept of possibilistic probability as a particularization of that of "possibilistic prevision". A possibilistic prevision is defined on a general set of "gambles" (numerical functions) instead of a set of events [73]. Its value $\tilde{P}(A)(p)$ for some probability value $p \in [0, 1]$ can be interpreted as the modeller's "upper betting rate" that the true probability for the event A is equal to p. This view makes sense if p represents a frequentist probability and $\tilde{P}(A)(p)$ is a subjective plausibility estimate.

A possibilistic probability $\tilde{P}$ is called *representable* [18] if there is a (second-order)[5] normal possibility distribution $\pi : \mathcal{P}_{\mathcal{A}'} \to [0, 1]$ that represents $\tilde{P}$, i.e., such that for all $p \in [0, 1]$ and $A \in \mathcal{A}'$, $\tilde{P}(A)(p) = \sup\{\pi(Q) : Q(A) = p\}$.

Remark 1 Zadeh [82] tries to model natural language statements such as "it is likely that John is short", where *likely* is the label of a fuzzy set of probabilities (that is, a possibility distribution μ_{likely} representing a fuzzy probability interval) and μ_{short} is the membership function of the linguistic term *short* on the height scale $[0, h]$. This piece of knowledge induces a second-order possibility distribution over the probability space. Namely, $\pi(Q)$ is the degree of membership of the expectation of "being short" to the fuzzy set of probabilities modeling *likely*, that is: $\pi(Q) = \mu_{likely}(Q(short))$ where $Q(short) = \int_0^h \mu_{short}(x)Q(x)dx$ is the expectation of the membership function μ_{short}, interpreted as the probability of the fuzzy event "John is short" [78].

In other words, in Zadeh's view, a higher-order possibility distribution is used to represent the meaning of a linguistic probability-qualified statement. In this chapter, the second-order possibility distribution π stems from a fuzzy random variable, and the fuzzy probability $\tilde{P}(A)$ is computed from π, not the other way around.

3.3 The Acceptability Function of a Fuzzy Random Variable

Consider two measurable spaces $(\Omega, \mathcal{A})$ and $(U, \mathcal{A}')$ and a fuzzy-valued mapping $\tilde{X} : \Omega \to \mathcal{F}(U)$. For each $\alpha \in [0, 1]$, we shall denote by $\tilde{X}_\alpha$ the α-cut of $\tilde{X}$, i.e., the multi-valued mapping $\tilde{X}_\alpha : \Omega \to \wp(U)$ that assigns, to each ω the α-cut of $\tilde{X}(\omega)$.

---

[5] The term "second-order" reflects that this possibility distribution is defined over a set of probability measures. In the general setting, the initial space is the class of linear previsions (including σ-additive probabilites as particular cases).

We shall say that $\tilde{X}$ is a *fuzzy random variable* (an FRV, for short) when every α-cut $\tilde{X}_\alpha$ is strongly measurable. This condition is equivalent to a classical measurability assumption: let us first consider, for each $A \in \mathcal{A}'$, the family of sets:

$$\mathcal{F}_A^\alpha = \{F \in \mathcal{F}(U) : F_\alpha \cap A \neq \emptyset\}.$$

Let us now denote by $\sigma_{\mathcal{F}}$ the σ-algebra generated by the class:

$$\{\mathcal{F}_A^\alpha : A \in \mathcal{A}', \alpha \in [0, 1]\} \subseteq \wp(\mathcal{F}(U)).$$

We easily check that a fuzzy-valued mapping is an FRV if and only if it is $\mathcal{A} - \sigma_{\mathcal{F}}$ measurable. We shall denote by $P \circ \tilde{X}^{-1}$ the probability measure induced by $\tilde{X}$ on $\sigma_{\mathcal{F}}$.

As we pointed out before, we shall consider the possibilistic interpretation of fuzzy sets and as a consequence, we shall follow Kruse and Meyer [53] interpretation of fuzzy random variables. According to it, an FRV represents the imprecise observation of the *original* random variable, $X_0 : \Omega \rightarrow U$. Thus, for each $\omega \in \Omega$ and each $x \in U$, the quantity $\tilde{X}(\omega)(x)$ represents the plausibility for x to be the "true" image of ω. (In other words, the plausibility that x coincides with $X_0(\omega)$).

According to this information, Kruse and Meyer define the degree of acceptability of an arbitrary measurable function, $X : \Omega \rightarrow U$, as the quantity

$$\mathrm{acc}_{\tilde{X}}(X) = \inf_{\omega \in \Omega} \tilde{X}(\omega)(X(\omega)).$$

For each $\omega \in \Omega$ and an arbitrary measurable mapping X, Kruse and Meyer interpret the membership value $\tilde{X}(\omega)(X(\omega))$ as the *acceptability* of the proposition "$X(\omega)$ coincides with $X_0(\omega)$". Furthermore, the infimum $\inf\{\tilde{X}(\omega)(X(\omega)) : \omega \in \Omega\}$ is intended as the acceptability of "$X(\omega) = X_o(\omega), \forall \omega \in \Omega$" or, in other words "$X$ coincides with X_0". Hence the quantity $\mathrm{acc}_{\tilde{X}}(X)$ is interpreted by the authors as the degree of acceptability of the proposition "X is the original variable". Next we give a further justification for the use of this infimum.

The function "$\mathrm{acc}_{\tilde{X}}$" takes values in the unit interval. In particular, so long as the fuzzy sets $\tilde{X}(\omega)$ are normal, there generally exists a choice of X such that $\mathrm{acc}_{\tilde{X}}(X) = 1$, namely, $\tilde{X}(\omega)(X(\omega)) = 1, \forall \omega \in \Omega$, that is, picking one value in the core of each $\tilde{X}(\omega)$.[6]

So $\mathrm{acc}_{\tilde{X}}$ can be formally regarded as a possibility distribution defined over the set of all $\mathcal{A} - \mathcal{A}'$ measurable functions from Ω to U, $\mathcal{X}_{\mathcal{A}-\mathcal{A}'}$.

The following result is an immediate consequence of the "possibilistic extension theorem" proved in [8] and also mentioned in [19] (it has been actually known earlier

---

[6] The existence of such a measurable mapping X is subject to some additional conditions. We can guarantee that X exists, for instance, if the α-levels of the images of $\tilde{X}$ are compact subsets of $\mathbb{R}^n$ or if the final space U is finite.

in the setting of possibilistic logic [28]).[7] We shall derive from it some interesting properties about the acceptability function.

Lemma 1 *Let U be a non-empty set and let C be a non-empty collection of subsets of U. Consider the set function $f : C \to [0, 1]$. Let us assume that, for every $\epsilon > 0$ there exists $x \in U$ such that $f(C) \geq 1 - \epsilon$, $\forall C \in C_x$, where $C_x = \{C \in C : x \in C\}$. Then, the greatest (least informative) possibility measure, $\Pi : \wp(U) \to [0, 1]$, which is dominated by f is the one associated to the possibility distribution $\pi : U \to [0, 1]$ defined as:*

$$\pi(x) = \begin{cases} \inf_{C \in C_x} f(C), & if \quad C_x \neq \emptyset \\ 1 & if \quad C_x = \emptyset. \end{cases}$$

If, furthermore, there exists a possibility measure, $\Pi' : \wp(U) \to [0, 1]$ such that $\Pi'(C) = f(C)$, $\forall C \in C$, then Π coincides with it on C.

Let us now consider the class of all possibility measures on $\wp(\mathcal{X}_{A-A'})$ that satisfy the following inequalities:

$$\Pi(\{X \in \mathcal{X}_{A-A'} : X(\omega) = x\}) \leq \tilde{X}(\omega)(x), \ \forall x \in U, \ \omega \in \Omega. \qquad (3.1)$$

According to last lemma, the acceptability function given by Kruse and Meyer is the possibility distribution that generates the greatest possibility measure in this class. Thus, the acceptability function is associated to the greatest (least informative) possibility measure, $\Pi_{\tilde{X}} : \wp(\mathcal{X}_{A-A'}) \to [0, 1]$, which is compatible with the available information about the images of X_0. Note that since the $\tilde{X}(\omega)$'s are normalized, so is the acceptability function.

The possibility distribution, $\mathrm{acc}_{\tilde{X}}$ is defined over the class of measurable functions from Ω to U, $\mathcal{X}_{A-A'}$, and it represents the information the fuzzy random variable provides about the "original" random variable. For an arbitrary class of measurable mappings, $\mathcal{X} \subseteq \mathcal{X}_{A-A'}$, the quantity

$$\Pi_{\tilde{X}}(\mathcal{X}) = \sup_{X \in \mathcal{X}} \inf_{\omega \in \Omega} \tilde{X}(\omega)(X(\omega))$$

represents the degree of plausibility that X_0 belongs to $\mathcal{X}$, denoting by $\Pi_{\tilde{X}}$ the associated to a possibility measure. Equivalently, the degree of certainty that X_0 belongs to $\mathcal{X}$ is

$$N_{\tilde{X}}(\mathcal{X}) = \inf_{X \notin \mathcal{X}} \sup_{\omega \in \Omega} 1 - \tilde{X}(\omega)(X(\omega)).$$

---

[7] In possibilistic logic, we use a collection of logical propositions with weights interpreted as confidence levels, viewed as lower bounds of necessity measures; it induces a minimally specific possibility distribution on interpretations of the language, that are elements of the sets of models of the propositions. If the set of propositions is consistent this possibility distribution is normal, but the corresponding necessity degrees may become greater than their original assignments. These updated necessity degrees satisfy a property similar to coherence in the sense of Walley.

Remark 2 When the fuzzy random variable is, in particular, a random set, Γ : $\Omega \to \wp(U)$, the acceptability function only takes the extreme values 0 and 1. In other words, it divides the class of measurable functions $\mathcal{X}_{A-A'}$ into two disjoint classes, $\mathcal{X}$ and $\mathcal{X}^c$, where $\mathcal{X}$ is, in fact, the class of measurable selections of Γ, i.e.:

$$\mathcal{X} = \{X \in \mathcal{X}_{A-A'} : \mathrm{acc}_\Gamma(X) = 1\}$$
$$= \{X \in \mathcal{X}_{A-A'} : X(\omega) \in \Gamma(\omega),\ \forall \omega \in \Omega\} = S(\Gamma).$$

Hence, our information about the original random variable can be described as follows: For each ω, all we know about $X_0(\omega)$ is that it belongs to the (crisp) set $\Gamma(\omega)$. In other words, all we know about the original random vector $X_0 : \Omega \to U$ is that it belongs to the class $S(\Gamma)$.

Now we further explain the acceptability function in terms of confidence levels. First, we give two lemmas.

Lemma 2 *The nested family of sets* $\{S(\tilde{X}_\alpha)\}_{\alpha \in [0,1]}$ *is a gradual set representing* $\mathrm{acc}_{\tilde{X}}(\tilde{X})$.

Proof Consider an arbitrary measurable mapping $X : \Omega \to U$. Then:

$$\mathrm{acc}_{\tilde{X}}(X) = \inf_{\omega \in \Omega} \tilde{X}(\omega)(X(\omega)) = \inf_{\omega \in \Omega} \sup\{\alpha \in [0, 1] : X(\omega) \in \tilde{X}_\alpha(\omega)\}$$
$$= \sup\{\alpha \in [0, 1] : X(\omega) \in \tilde{X}_\alpha(\omega),\ \forall \omega \in \Omega\} = \sup\{\alpha \in [0, 1] : X \in S(\tilde{X}_\alpha)\}.$$

$\square$

This representation of $\mathrm{acc}_{\tilde{X}}$ as the fuzzy set associated to the nested family $\{S(\tilde{X}_\alpha)\}_{\alpha \in [0,1]}$ is quite related to the interpretation of fuzzy sets as families of confidence sets, which is studied in detail in [14, 32]. Let us briefly explain these ideas. First, we can easily check the following result related to Theorem 1:

Lemma 3 *Let us consider a nested family of subsets,* $\{A_\alpha\}_{\alpha \in [0,1]}$, *of a universe U. The class of probability measures that satisfy the inequalities*

$$P(A_\alpha) \geq 1 - \alpha,\ \forall \alpha \in [0, 1]$$

coincides with the class of probability measures dominated by the possibility measure Π :

$$\Pi(C) = \sup\{\alpha \in [0, 1] : C \cap A_\alpha \neq \emptyset\},\ \forall C \subseteq U.$$

(or equivalently, dominating the necessity measure $N(C) = \inf\{1 - \alpha : A_\alpha \subseteq C\}$).

Based on this result, we can give $\mathrm{acc}_{\tilde{X}}$ a "confidence degree" interpretation. Let us suppose that an expert provides the following information about the images of $X_0 : \Omega \to \mathbb{R}$:

$\forall \omega \in \Omega, \forall \alpha \in (0, 1]$, $X_0(\omega)$ *belongs to* $[\tilde{X}(\omega)]_\alpha$ *with probability greater than or equal to* $1 - \alpha$.

According to Lemma 3, we observe that $\mathrm{acc}_{\tilde{X}}$ represents this information. In other words, the expert asserts that X_0 is a selection of $\tilde{X}_\alpha$ with probability greater than or equal to $1 - \alpha$ for every $\alpha \in (0, 1]$. Note that indeed $N_{\tilde{X}}([\tilde{X}(\omega)]_\alpha) \geq 1 - \alpha$. So, the class of probability measures (defined on a class of $\mathcal{A} - \mathcal{A}'$ measurable functions[8]) that satisfy these inequalities coincides with the class of probability measures that are dominated by $\Pi_{\tilde{X}}$ (the possibility measure associated to $\mathrm{acc}_{\tilde{X}}$).

3.4 Second-Order Possibility Measures Associated to a Fuzzy Random Variable

Based on this interpretation of a fuzzy random variable as an imprecise observation of the "original" variable, we shall define three different possibility distributions over the set of all probability measures on $\mathcal{A}'$. We shall denote them by $\pi_{\tilde{X}}$, $\pi_{\tilde{X}}^{\downarrow}$ and $\pi_{\tilde{X}}^{*}$, respectively. Under the above interpretation of fuzzy random variables, the first one of them will represent the available (imprecise) information about the *original* probability measure.[9] The other two possibility measures will also arise in a natural way, but they are less informative. In particular, $\pi_{\tilde{X}}^{\downarrow}$ will be associated to "fuzzy-valued probabilities" defined on $\mathcal{A}'$. On the other hand, $\pi_{\tilde{X}}^{*}$ will be related to the family of pairs of Dempster upper and lower probabilities associated to the α-cuts of $\tilde{X}$. The procedure described below is briefly outlined in [15].

3.4.1 The Fuzzy Probability Envelope

Definition 2 Let us consider a probability space $(\Omega, \mathcal{A}, P)$, a measurable space $(U, \mathcal{A}')$, and a fuzzy random variable $\tilde{X} : \Omega \to \mathcal{F}(U)$ and the class $\mathcal{P}_{\mathcal{A}'}$ of all (σ-additive) probability measures over $\mathcal{A}'$. The *fuzzy probability envelope associated to* $\tilde{X}$, $\pi_{\tilde{X}} : \mathcal{P}_{\mathcal{A}'} \to [0, 1]$, is defined as:

$$\pi_{\tilde{X}}(Q) = \sup\{\mathrm{acc}_{\tilde{X}}(X) : P_X = Q\}, \ \forall Q \in \mathcal{P}_{\mathcal{A}'}.$$

Under the last conditions, we shall denote by $\Pi_{\tilde{X}}$ (resp. $N_{\tilde{X}}$) the possibility (resp. necessity) measure associated to the fuzzy probability envelope:

$$\Pi_{\tilde{X}}(\mathcal{P}) = \sup\{\mathrm{acc}_{\tilde{X}}(X) : P_X \in \mathcal{P}\}, \ \forall \mathcal{P} \subseteq \mathcal{P}_{\mathcal{A}'}.$$

$$N_{\tilde{X}}(\mathcal{P}) = \inf\{1 - \mathrm{acc}_{\tilde{X}}(X) : P_X \notin \mathcal{P}\}, \ \forall \mathcal{P} \subseteq \mathcal{P}_{\mathcal{A}'}.$$

---

[8] We can consider the σ-algebra generated by the nested family of classes $\{S(\tilde{X}_\alpha)\}_{\alpha \in [0,1]}$.

[9] From now on, we shall call "original" the probability measure $P_{X_0} = P \circ X_0^{-1}$ associated to the original random variable.

It represents the imprecise information that the FRV provides about the original probability measure. For an arbitrary class of probability measures, $\Pi_{\tilde{X}}(\mathcal{P})$ (resp. $N_{\tilde{X}}(\mathcal{P})$) quantifies the plausibility (resp. certainty) that the original probability belongs to the class $\mathcal{P}$, under the available information: let us observe that they can be written as $\Pi_{\tilde{X}}(\mathcal{X}(\mathcal{P}))$, and $N_{\tilde{X}}(\mathcal{X}(\mathcal{P}))$, where $\mathcal{X}(\mathcal{P})$ denotes the class of random variables whose associated probability measure belongs to $\mathcal{P}$.

$\Pi_{\tilde{X}}$ is a second-order possibility measure in the sense of [18, 20], as it is a possibility measure defined on a class of linear previsions. In our particular model, only σ-additive probabilities have non-zero values of possibility.

Furthermore, the concept of fuzzy probability envelope is closely related to the definition of probability envelope recalled in Chap. 1. In fact, the (nested) family of sets of probability measures $\{\mathcal{P}(\tilde{X}_\alpha)\}_{\alpha \in [0,1]}$ is a gradual set representation of $\pi_{\tilde{X}}$ as we shall prove next.

Proposition 1 $\{\mathcal{P}(\tilde{X}_\alpha)\}_{\alpha \in [0,1]}$ *is a gradual set representation of* $\pi_{\tilde{X}}$.

Proof Consider an arbitrary probability measure, Q. By Lemma 2, we know that $\pi_{\tilde{X}}(Q)$ coincides with the quantity:

$$\sup\{\alpha : \exists X \in S(\tilde{X}_\alpha) \text{ s.t. } Q = P_X\}.$$

On the other hand, by definition of $\{\mathcal{P}(\tilde{X}_\alpha)\}_{\alpha \in [0,1]}$ the class $\{\alpha : \exists X \in S(\tilde{X}_\alpha) \text{ s.t. } Q = P_X\}$ coincides with $\{\alpha : Q \in \mathcal{P}(\tilde{X}_\alpha)\}$. Hence, we deduce that $\pi_{\tilde{X}}(Q)$ coincides with $\sup\{\alpha : Q \in \mathcal{P}(\tilde{X}_\alpha)\}$. In words, $\{\mathcal{P}(\tilde{X}_\alpha)\}_{\alpha \in [0,1]}$ is a gradual set representation of $\pi_{\tilde{X}}$. $\square$

We deduce from this result that the concept of "fuzzy probability envelope" is closely related to the concept of "expectation of a fuzzy random variable" introduced by Puri and Ralescu in [62]. In fact, the fuzzy expectation of an FRV [62] is defined as the fuzzy set associated to the nested family $\{E(\tilde{X}_\alpha)\}_{\alpha \in [0,1]}$, where $E(\tilde{X}_\alpha) = \{\int X\,dP : X \in S(\tilde{X}_\alpha)\}$ represents the Aumann expectation of $\tilde{X}_\alpha$.

Furthermore, this representation of $\pi_{\tilde{X}}$ as the fuzzy set associated to the nested family $\{\mathcal{P}(\tilde{X}_\alpha)\}_{\alpha \in [0,1]}$ is quite related to the interpretation of fuzzy sets as families of confidence sets [14] outlined in Sect. 3.2.1. According to Lemma 3, we can give $\pi_{\tilde{X}}$ a "confidence degree" interpretation. Let us suppose that an expert provides the following information about the images of $X_0 : \Omega \to \mathbb{R}$:

For each $\alpha \in [0, 1]$, $X_0(\omega)$ *belongs to* $[\tilde{X}(\omega)]_\alpha$, $\forall \omega \in \Omega$, *with probability greater than or equal to* $1 - \alpha$.

In other words, the expert asserts that X_0 is a selection of $\tilde{X}_\alpha$ with probability greater than or equal to $1 - \alpha$. Under these conditions, we can represent the available (imprecise) information about the original probability measure as follows:

$\forall \alpha \in [0, 1]$, P_{X_0} *belongs to* $\mathcal{P}(\tilde{X}_\alpha)$ *with probability greater than or equal to* $1 - \alpha$.

In other words, the available information about P_{X_0} is represented by the class of second-order probability measures $\mathbb{P}$ such that:

$$\mathbb{P}(\mathcal{P}(\tilde{X}_\alpha)) \geq 1 - \alpha, \forall \alpha \in [0, 1] \tag{3.2}$$

This statement is to be viewed in the light of the obvious inequality $I\!N_{\tilde{X}}(\mathcal{P}(\tilde{X}_\alpha)) \geq 1 - \alpha$. According to Lemma 3, we observe that $\pi_{\tilde{X}}$ determines the same information as the set of probability measures satisfying Eq. (3.2), i.e., the class of second-order probabilities that satisfy these inequalities coincides with the class of second-order probabilities that are dominated by $I\!I_{\tilde{X}}$ (the possibility measure associated to $\pi_{\tilde{X}}$).

3.4.2 The Fuzzy Probability of Events Induced by a Fuzzy Random Variable

Based on the interpretation of $\mathrm{acc}_{\tilde{X}}$ and $\Pi_{\tilde{X}}$, we can also define in a natural way the "fuzzy-valued probability measure" of $\tilde{X}$ as a mapping, $\tilde{P}_{\tilde{X}}$, that assigns a fuzzy subset of the unit interval to each particular event $A \in \mathcal{A}'$.

Definition 3 Let us consider a probability space $(\Omega, \mathcal{A}, P)$, a measurable space $(U, \mathcal{A}')$, and a fuzzy random variable $\tilde{X} : \Omega \to \mathbb{R}^n$. The *fuzzy probability measure*[10] *associated to* $\tilde{X}$, $\tilde{P}_{\tilde{X}} : \mathcal{A}' \to \mathcal{F}([0, 1])$, will be defined as:

$$\tilde{P}_{\tilde{X}}(A)(p) = \sup\{\mathrm{acc}_{\tilde{X}}(X) : P_X(A) = p\}, \ \forall p \in [0, 1], \ \forall A \in \mathcal{A}'.$$

For an arbitrary event $A \in \mathcal{A}'$, the fuzzy value $\tilde{P}_{\tilde{X}}(A)$ represents, in this framework, the imprecise information the FRV provides about the true probability of A, $P_{X_0}(A)$. Specifically, $\tilde{P}_{\tilde{X}}(A)(p)$ represents the degree of possibility that $P_{X_0}(A)$ coincides with p. From a formal point of view, $\tilde{P}_{\tilde{X}}$ is a possibilistic probability in the sense of Sect. 3.2.2 so we can ask ourselves whether it is "representable" and what is its least informative representation (greatest second-order possibility distribution). We can easily check that $\tilde{P}_{\tilde{X}}(A)$ can be expressed as:

$$\tilde{P}_{\tilde{X}}(A)(p) = \sup\{\pi_{\tilde{X}}(Q) : Q \text{ prob. with } Q(A) = p\}.$$

In other words, $\tilde{P}_{\tilde{X}}$ is, in fact, representable, and $\pi_{\tilde{X}}$ is a possible representation. But it is not, in general, its least specific one. Based on the formulae given in [18], we observe that the least specific possibilistic representation of $\tilde{P}_{\tilde{X}}$ is the second-order possibility distribution $\pi_{\tilde{X}}^{\downarrow}$, defined as:

---

[10] The term "fuzzy probability measure" is used here in a very loose sense. In fact, $\tilde{P}_{\tilde{X}}$ does not satisfy in general the additivity property (the main property of probability measures).

$$\pi_{\tilde{X}}^{\downarrow}(Q) = \inf\{\tilde{P}_{\tilde{X}}(A)(Q(A)) : A \in \mathcal{A}'\}, \ \forall Q \in \mathcal{P}_{\mathcal{A}'}.$$

Let us now consider, for each α, the collection of probability measures (already introduced in Sect. 2.3.1 of the last chapter):

$$\mathcal{P}^{\downarrow}(\tilde{X}_{\alpha}) = \{Q : \ Q(A) \in \mathcal{P}(\tilde{X}_{\alpha})(A), \forall A\}.$$

It contains the probability envelope $\mathcal{P}(\tilde{X}_{\alpha})$, but they do not coincide in general, as we shall illustrate in Examples 5 and 6. Next we check that the (nested) family of classes of probability measures $\{\mathcal{P}^{\downarrow}(\tilde{X}_{\alpha})\}_{\alpha\in[0,1]}$ is a gradual set associated to the fuzzy set $\pi_{\tilde{X}}^{\downarrow}$.

Proposition 2 $\{\mathcal{P}^{\downarrow}(\tilde{X}_{\alpha})\}_{\alpha\in[0,1]}$ *is a gradual set representation of* $\pi_{\tilde{X}}^{\downarrow}$.

Proof First of all, let us consider, for each α and each $A \in \mathcal{A}'$, the set of values $\mathcal{P}(\tilde{X}_{\alpha})(A) := \{Q(A) : \ Q \in \mathcal{P}(\tilde{X}_{\alpha})\}$. It is easily checked that the family $\{\mathcal{P}(\tilde{X}_{\alpha})(A)\}_{\alpha\in[0,1]}$ is a gradual set representation of $P_{\tilde{X}}(A)$, for each $A \in \mathcal{A}'$. In other words, for any $A \in \mathcal{A}'$ and Q, the following equality holds:

$$\tilde{P}_{\tilde{X}}(A)(Q(A)) = \sup\{\alpha \in [0, 1] : Q(A) \in \mathcal{P}(\tilde{X}_{\alpha})(A)\}.$$

On the other hand, we observe that the quantity $\inf_{A\in\mathcal{A}'} \sup\{\alpha \in [0, 1] : Q(A) \in \mathcal{P}(\tilde{X}_{\alpha})(A)\}$ can be written as $\sup\{\alpha \in [0, 1] : Q(A) \in \mathcal{P}(\tilde{X}_{\alpha})(A), \forall A \in \mathcal{A}'\}$. We deduce that

$$\pi_{\tilde{X}}^{\downarrow}(Q) = \sup\{\alpha \in [0, 1] : Q(A) \in \mathcal{P}(\tilde{X}_{\alpha})(A), \forall A \in \mathcal{A}'\} = \sup\{\alpha \in [0, 1] : Q \in \mathcal{P}^{\downarrow}(\tilde{X}_{\alpha})\}.$$
□

Let us now compare the information provided by $\pi_{\tilde{X}}$ and the information provided by $\tilde{P}_{\tilde{X}}$. In other words, let us compare $\pi_{\tilde{X}}$ and $\pi_{\tilde{X}}^{\downarrow}$. On the one hand, as we checked above, the fuzzy probability measure associated to $\tilde{X}$ is representable [18] by the (normal) possibility distribution $\pi_{\tilde{X}}$. On the other hand, its greatest representation is $\pi_{\tilde{X}}^{\downarrow}$. Then we derive that

$$\pi_{\tilde{X}}(Q) \leq \pi_{\tilde{X}}^{\downarrow}(Q), \ \forall Q.$$

In other words,

$$\Pi_{\tilde{X}}(\mathcal{P}) \leq \Pi_{\tilde{X}}^{\downarrow}(\mathcal{P}), \ \forall \mathcal{P} \tag{3.3}$$

Let us now remind that any possibility measure is equivalent to the class of probability measures it dominates. Hence, the information provided by $\Pi_{\tilde{X}}$ and $\Pi_{\tilde{X}}^{\downarrow}$ can be represented by individual classes of second-order probability measures. Based on Eq. (3.3), the class of second-order probability measures associated to $\Pi_{\tilde{X}}$ is included in the class (of second-order probabilities) associated to $\Pi_{\tilde{X}}^{\downarrow}$. This means that the possibility distribution $\pi_{\tilde{X}}$ (or, equivalently, the possibility measure $\Pi_{\tilde{X}}$) provides at least as much information about the "original" probability measure as the fuzzy probability measure does. The second-order possibility distributions $\pi_{\tilde{X}}$ and $\pi_{\tilde{X}}^{\downarrow}$ do not coincide in general, as we shall show in Example 2.

3.4.3 The Fuzzy Credal Set Induced by an FRV

We shall deal now with the third and last second-order possibility distribution associated to $\tilde{X}$, $\pi_{\tilde{X}}^{*}$.

Definition 4 Let us consider a probability space $(\Omega, \mathcal{A}, P)$, a measurable space $(U, \mathcal{A}')$, and a fuzzy random variable $\tilde{X} : \Omega \to \mathcal{F}(U)$. Let us consider the class $\mathcal{P}_{\mathcal{A}'}$ of all (σ-additive) probability measures over $\mathcal{A}'$. The *convex fuzzy credal set* is defined as the possibility distribution associated to the gradual set $\{\mathcal{M}(P_{\tilde{X}_{\alpha}}^{*})\}_{\alpha \in [0,1]}$. In other words, $\pi_{\tilde{X}}^{*}$ is a possibility distribution over probability measures Q over $\mathcal{A}'$, defined as

$$\pi_{\tilde{X}}^{*}(Q) = \sup\{\alpha \in [0,1] : Q(A) \leq P_{\tilde{X}_{\alpha}}^{*}(A), \ \forall A \in \mathcal{A}', \ \forall Q \in \mathcal{P}_{\mathcal{A}'}\}.$$

It can be called the fuzzy credal set of $\tilde{X}$, since it is the generalisation of the convex envelope to FRV's. According to Propositions 1 and 2, the second-order possibility measure $\pi_{\tilde{X}}^{*}$ dominates both $\pi_{\tilde{X}}$ and $\pi_{\tilde{X}}^{\downarrow}$, i.e.:

$$\pi_{\tilde{X}}(Q) \leq \pi_{\tilde{X}}^{\downarrow}(Q) \leq \pi_{\tilde{X}}^{*}(Q), \ \forall Q \in \mathcal{P}_{\mathcal{A}'}. \tag{3.4}$$

In Fig. 3.1 we illustrate the dominance relations among these second-order possibility distributions. We recall the notation we use for each one of them when $\tilde{X}$ is, in particular, a random set.

It is interesting to bridge the gap between the fuzzy probability distributions and Dempster upper and lower probabilities. Namely, $\pi_{\tilde{X}}^{*}(A)$ is the fuzzy extension of the interval $[P_{*}(A), P^{*}(A)]$ in the sense that the lower and upper bounds of each α-cut $\pi_{\tilde{X}}^{*}(A)_{\alpha} = [P_{*\alpha}(A), P_{\alpha}^{*}(A)]$ are respectively Dempster lower and upper probabilities associated to the multivalued mapping $\tilde{X}_{\alpha} : \Omega \to \wp(U)$, that associates, to each $\omega \in \Omega$, the α-level of the fuzzy set $\tilde{X}(\omega)$. In other words, if we consider the cut $\pi_{\tilde{X}}^{*}(A)_{\alpha}$ its lower bound is infinite-monotone and its upper bound is infinite alternating. So, the lower and upper probabilities become gradual numbers [35]

Fig. 3.1 Dominance relations between different second-order possibility distributions

less information →

f.r.v.	$\pi_{\tilde{X}}$	$\pi_{\tilde{X}}^{\downarrow}$	$\pi_{\tilde{X}}^{*}$
random set	$\mathcal{P}(\Gamma)$	$\mathcal{P}^{\downarrow}(\Gamma)$	$\mathcal{M}(P^{*})$

$\alpha \mapsto P_{*\alpha}$ and $\alpha \mapsto P_\alpha^*$ that can be called upper and lower fuzzy probability functions. It is thus clear that there is no such thing as an additive fuzzy (nor set-valued) probability measure induced by a fuzzy probability assignment or density. Nevertheless, it is possible to express a generalisation of the additivity property due to the infinite-monotonicity of the lower fuzzy probability and the infinite-alternation of the upper fuzzy probability functions, in the form of a fuzzy set inclusion

$$\pi_{\tilde{X}}^*(A \cup B) \oplus \pi_{\tilde{X}}^*(A \cap B) \subseteq \pi_{\tilde{X}}^*(A) \oplus \pi_{\tilde{X}}^*(B).$$

where $\oplus$ is the usual fuzzy number addition. It cannot be written at any order n, since the order-n alternating properties of upper probabilities provides an upper bound to $P(\cap_{i=1}^n A_i)$, not to $P(\cup_{i=1}^n A_i)$ (see Sect. 2.2.3). And the expression at order 2 does not generally imply its counterpart at order $n > 2$. However, according to the super-additivity of the lower bound, and the sub-additivity of the upper bound of every cut, we also have (for disjoint sets and $n \geq 2$), the inclusion $\pi_{\tilde{X}}^*(\cup_{i=1}^n A_i) \subseteq$ $\oplus_{i=1}^n \pi_{\tilde{X}}^*(A_i)$. But even if similar to additivity (and reducing to it if the fuzzy set-valued probabilities are singletons), these expressions cannot be handled as in the usual case: there is no way of moving any term from one side of the inclusion to the other while changing the sign in front of the term.

Remark 3 (**FRV's and belief functions**) Random fuzzy sets can be the basis for a generalized theory of fuzzy belief functions. Namely, in the finite case if $\{(\tilde{F}_i, p_i) : i = 1, \ldots, n\}$ represents the probability masses associated to the images $\tilde{X}(\omega)$ when varying ω, it can be viewed as defining a set of fuzzy focal elements. The degree of belief in a fuzzy event, based on fuzzy focal elements can be defined from two points of view:

- the first one is to exploit degrees of comparison between fuzzy sets: degree of inclusion $I(\tilde{F}_i, \tilde{F})$ for the belief function, and degree of overlap $O(\tilde{F}_i, \tilde{F})$ for the plausibility function (starting with Zadeh [80], then Yager [75], Ishizuka et al. [48], Dubois and Prade [29]). Namely

$$Bel(\tilde{F}) = \sum_{i=1,\ldots,n} p_i I(\tilde{F}_i, \tilde{F}); \quad Pl(\tilde{F}) = \sum_{i=1,\ldots,n} p_i O(\tilde{F}_i, \tilde{F}).$$

If $\tilde{F}$ is a crisp set A, and π_i is the possibility distribution for $\tilde{F}_i$, we can define $I(\tilde{F}_i, A) = N_i(A)$ and $O(\tilde{F}_i, A) = \Pi_i(A)$. Then, $[Bel(A), Pl(A)]$ is the mean value interval [30] of $\tilde{P}_{\tilde{X}}(A)$, namely $Bel(A) = \int_0^1 \inf \tilde{P}_{\tilde{X}}(A)_\alpha d\alpha$ and $Pl(A) = \int_0^1 \sup \tilde{P}_{\tilde{X}}(A)_\alpha d\alpha$ [7].
- The second approach after Yen [76, 77] extends Smets [71]'s definition of the degree of belief in a fuzzy event based on a Choquet integral, now considering fuzzy focal elements as convex combinations of sets, i.e., $\tilde{F}_i = \{(\tilde{F}_i^j, q_i^j) : j = 1, \ldots, m\}$. It gives

$$Bel(\tilde{F}) = \sum_{i=1,\ldots,n} p_i \sum_{j=1,\ldots,m} q_i^j \min_{u \in \tilde{F}_i^j} \tilde{F}(u).$$

$$Pl(\tilde{F}) = \sum_{i=1,\ldots,n} p_i \sum_{j=1,\ldots,m} q_i^j \max_{u \in \tilde{F}_i^j} \tilde{F}(u).$$

The two approaches coincide if $\tilde{F}$ is a crisp set. Another work along this line appears in [21].

3.4.4 Statistical Parameters Under the Epistemic Approach

The procedure that computes set-valued statistical parameters (see Sect. 2.4 of Chap. 1) can be generalized to fuzzy random variables. For a particular parameter $\theta(P_{X_0})$, the fuzzy set $\theta(\pi_{\tilde{X}})$ defined as

$$\theta(\pi_{\tilde{X}})(x) = \sup\{acc_{\tilde{X}}(X) : \theta(P_X) = x\} = \sup\{\pi_{\tilde{X}}(Q) : \theta(Q) = x\}$$

represents the available imprecise information about $\theta(P_{X_0})$. Puri and Ralescu expectation [62], second-order variance [11, 12] and the fuzzy probability measure are compatible with this definition. Under the above interpretation of fuzzy random variables, $\theta(P_X)(x)$ represents the degree of plausibility that x coincides with $\theta(P_{X_0})$.

In the following section, we shall compare the information provided by $\pi_{\tilde{X}}, \pi_{\tilde{X}}^{\downarrow}$, $\pi_{\tilde{X}}^*$ and $P \circ \tilde{X}^{-1}$ about different aspects of P_{X_0} and check their impact on statistical parameters. In particular, we shall deal with Shannon entropy (Examples 5 and 6), variance (Examples 7 and 8).

3.5 Discrepancies Between Fuzzy Random Variables and Their Associated Fuzzy Credal Sets

As we have stated in the previous section, a fuzzy random variable represents here the imprecise observation of a (classical) random variable $X_0 : \Omega \to \mathbb{R}^n$. Each image $\tilde{X}(\omega)$ is a possibility distribution on $\mathbb{R}^n$ that represents the available information about the point $X_0(\omega)$. In this setting, the second-order possibility measure $\pi_{\tilde{X}}$ arises in a natural way to represent the available information about P_{X_0}. Our main purpose in this section is to show that the "fuzzy probability envelope" associated to an FRV is a good representation of this information, but the probability measure induced by $\tilde{X}$ on $\sigma_{\mathcal{F}}$ is not. In fact, we can find two different fuzzy random variables inducing the same (classical) probability measure but with different "fuzzy probability envelopes".

In this section, we provide counterexamples showing that the three possibility distributions $\pi_{\tilde{X}}(Q), \pi_{\tilde{X}}^{\downarrow}(Q), \pi_{\tilde{X}}^*(Q)$ induced by a fuzzy random variable may be different. We shall first deal with the particular case of random sets (Examples 5, 7 and 9), and then, we shall derive some conclusions about the general case of fuzzy random variables (Examples 6, 8 and 10). We also consider the probability measure associated to a pair of independent identically distributed random variables (Examples 9 and 10).

3.5.1 Probability Envelopes Versus Credal Sets: The Case of Random Sets

So, let us now study the particular case of random sets (fuzzy random variables whose images are crisp sets) and illustrate the differences among the sets of probabilities $\mathcal{P}(\Gamma)$, $\mathcal{P}^{\downarrow}(\Gamma)$ and $\mathcal{M}(P_{\Gamma}^{*})$ introduced in the previous chapter. We shall then return to the general case (FRV), and obtain some conclusions about the differences among $\pi_{\tilde{X}}$, $\pi_{\tilde{X}}^{\downarrow}$ and $\pi_{\tilde{X}}^{*}$.

Let us consider an arbitrary random set $\Gamma : \Omega \to \wp(U)$. It was pointed out in the previous chapter that the following relations hold:

$$\mathcal{P}(\Gamma) \subseteq \mathcal{P}^{\downarrow}(\Gamma) \subseteq \mathcal{M}(P_{\Gamma}^{*}).$$

Furthermore, the information provided by $\mathcal{M}(P_{\Gamma}^{*})$ is determined by the upper probability function P_{Γ}^{*}. In most practical situations,[11] the upper probability $P_{\Gamma}^{*}(A)$ coincides with the supremum:

$$\sup\{P_X(A) : X \in S(\Gamma)\} = \sup\{Q(A) : Q \in \mathcal{P}(\Gamma)\} = \sup \mathcal{P}(\Gamma)(A).$$

(See [58, 59] for a detailed study.) Nevertheless, recall that the classes of probability measures $\mathcal{P}(\Gamma)$, $\mathcal{P}^{\downarrow}(\Gamma)$ and $\mathcal{M}(P_{\Gamma}^{*})$ may be quite different, even in the situations where both quantities $P_{\Gamma}^{*}(A)$ and $\sup \mathcal{P}(\Gamma)(A)$ do coincide for every A. These differences are reflected on the information about statistical parameters like the entropy or the variance, as we shall respectively show in Examples 5 and 7.

Example 5 Let Ω be a universe with only two elements, $\Omega = \{\omega_1, \omega_2\}$. Let us consider the probability measure $P : \wp(\Omega) \to [0, 1]$ such that $P(\{\omega_1\}) = 2/3$ (and $P(\{\omega_2\}) = 1/3$). Let us define the random set $\Gamma : \Omega \to \wp(\mathbb{R})$ as follows:

$$\Gamma(\omega_1) = \{0, k\} \text{ and } \Gamma(\omega_2) = \{-k, 0, k\}.$$

---

[11] As pointed out in Chap. 1, those situations include the cases where the final set U is finite or the images of Γ are compact subsets of $\mathbb{R}^n$.

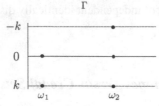

Let us identify each probability measure, Q, on the final space with the triplet $(Q(\{-k\}), Q(\{0\}), Q(\{k\}))$, and let us calculate the expressions for $\mathcal{P}(\Gamma)$, $\mathcal{P}^{\downarrow}(\Gamma)$ and $\mathcal{M}(P_{\Gamma}^{*})$.

- Calculation of $\mathcal{P}(\Gamma)$:
  $\mathcal{P}(\Gamma) = \{(1/3, 2/3, 0), (0, 1, 0), (0, 2/3, 1/3), (1/3, 0, 2/3), (0, 1/3, 2/3), (0, 0, 1)\}$
  As Ω contains two elements, the support of the original probability distribution on $\{-k, 0, k\}$ induced by Γ has no more than 2 elements as well.
- Calculation of $\mathcal{P}^{\downarrow}(\Gamma)$:
  Prior to obtaining $\mathcal{P}^{\downarrow}(\Gamma)$, we first need to calculate $\mathcal{P}(\Gamma)(A)$, for each $A \subseteq \{-k, 0, k\}$.
  $\mathcal{P}(\Gamma)(\{0\}) = \mathcal{P}(\Gamma)(\{k\}) = \{0, 1/3, 2/3, 1\}$,
  $\mathcal{P}(\Gamma)(\{-k\}) = \{0, 1/3\}$,
  $\mathcal{P}(\Gamma)(\{-k, 0\}) = \{0, 1/3, 2/3, 1\}$,
  $\mathcal{P}(\Gamma)(\{-k, k\}) = \{0, 1/3, 2/3, 1\}$,
  $\mathcal{P}(\Gamma)(\{0, k\}) = \{2/3, 1\}$, and
  $\mathcal{P}(\Gamma)(\{-k, 0, k\}) = \{1\}$.
  Hence, we deduce that:

$$\mathcal{P}^{\downarrow}(\Gamma) = \mathcal{P}(\Gamma) \cup \{(1/3, 1/3, 1/3)\}.$$

As already said in Sect. 2.3.1, the above set is only based on the set-valued probabilities of event, obtained by a projection. The uniform distribution is consistent with these set-valued probabilities, not with the original random set.
- Calculation of $\mathcal{M}(P_{\Gamma}^{*})$:
  Prior to the calculation of $\mathcal{M}(P_{\Gamma}^{*})$, we first need to calculate the expression of P_{Γ}^{*} (which is nothing but a plausibility function in the sense of Shafer, based on mass assignment $m(\{-k, 0, k\}) = 1/3$, $m(\{0, k\}) = 2/3$. We have:

$$P_{\Gamma}^{*}(A) = P(\{\omega \in \Omega : \Gamma(\omega) \cap A \neq \emptyset\})$$
$$= \begin{cases} 1 & \text{if } A \cap \{0, k\} \neq \emptyset, \\ 1/3 & \text{if } A \cap \{0, k\} = \emptyset, \text{and} - k \in A \\ 0 & \text{if } A = \emptyset, \end{cases}$$

Hence, we observe that $\mathcal{M}(P_{\Gamma}^{*}) = \{(p_1, p_2, p_3) : p_1 \leq 1/3\}$. In fact, $\mathcal{M}(P_{\Gamma}^{*})$ is the convex hull of $\mathcal{P}(\Gamma)$. Then it is clear that

$$\mathcal{P}(\Gamma) \subsetneq \mathcal{P}^{\downarrow}(\Gamma) \subsetneq \mathcal{M}(P_{\Gamma}^{*}) = \text{Conv}(\mathcal{P}(\Gamma)).$$

Thus, if we need to calculate the tightest upper and lower bounds for the true probability of an event, the differences among $\mathcal{P}(\Gamma), \mathcal{P}^{\downarrow}(\Gamma)$ and $\mathcal{M}(P_\Gamma^*)$ are not relevant.

3.5.2 Ill-Known Entropy

Nevertheless, when we replace $\mathcal{P}(\Gamma)$ by $\mathcal{P}^{\downarrow}(\Gamma)$ or $\mathcal{M}(P_\Gamma^*)$ we lose important information about other properties of the original probability, P_{X_0}. For instance, the difference between $\mathcal{P}(\Gamma)$ and $\mathcal{P}^{\downarrow}(\Gamma)$ affects the calculus of Shannon's entropy. $\mathcal{P}(\Gamma)$ provides much more information than $\mathcal{P}^{\downarrow}(\Gamma)$ about the entropy $H(p) = -\sum_{i=1,n} p_i \log_2 p_i$ of the original probability measure, as we show below. For an arbitrary class of probability measures $\mathcal{P}$, let us denote by $H(\mathcal{P})$ the set of values of entropy for probability measures in $\mathcal{P}$:

$$H(\mathcal{P}) = \{H(Q) : Q \in \mathcal{P}\}.$$

Example 5 (continued) We easily check that:

$$H(\mathcal{P}(\Gamma)) = \{0, \log_2 3 - 2/3\},$$
$$H(\mathcal{P}^{\downarrow}(\Gamma)) = \{0, \log_2 3 - 2/3, \log_2 3\} \text{ and}$$
$$H(\mathcal{M}(P_\Gamma^*)) = [0, \log_2 3].$$

Let us summarize the conclusions we derive from this example. We have considered a random experiment whose possible results are ω_1 and ω_2, with respective probabilities $2/3$ and $1/3$. The random variable $X_0 : \Omega \to \mathbb{R}$ represents a certain attribute of these two outcomes. But this attribute is imprecisely observed. In fact, all we know about $X_0(\omega_i)$ is that it belongs to the set of values $\Gamma(\omega_i)$, $i = 1, 2$. Hence, we know that P_{X_0} is dominated by P_Γ^*, i.e., $P_{X_0} \in \mathcal{M}(P_\Gamma^*)$. We would conclude that the uncertainty about the image of X_0 (the entropy of X_0) is less than $\log_2 3$ bit. But this is not the end of it. We not only know that P_{X_0} is dominated by P_Γ^*, but we also know that it belongs to $\mathcal{P}(\Gamma)$. Based on this (more precise) information, we realize that the entropy of X_0 cannot be greater than $(\log_2 3 - 2/3)$ bit.

We see that the information provided by $\mathcal{P}(\Gamma)$ about P_{X_0} is, in general, more precise than the information determined by the upper probability function P_Γ^*. We have shown here that these differences are reflected in the information about the entropy. Such differences would also appear with other statistical parameters like variance. Even the expectation can be affected as the set of expectations $\mathcal{E}(\mathcal{P}(\Gamma))$ of probabilities in $\mathcal{P}(\Gamma)$ will differ from the interval bounded by the upper and lower expectations induced by P^* and P_*, even if these upper and lower expectations will be contained in $\mathcal{E}(\mathcal{P}(\Gamma))$ (contrary to the case of entropy here).

Notice that P_Γ^* determines the probability distribution of Γ (regarded as a $\wp(\Omega) - \wp(\wp(\{-k, 0, k\}))$ measurable function). Hence we derive that the probability distribution of Γ, $P \circ \Gamma^{-1}$, (based on the mass assignment in the sense of

Shafer) does not provide all the relevant information when Γ represents the imprecise observation of a random variable.

Remark 4 We can slightly modify the conditions in the last example and define another random set with the same probability distribution as Γ, but a different probability envelope. Let us consider the initial space $([0, 1], \beta_{[0,1]}, \lambda_{[0,1]})$ (the unit interval, with the usual Borel σ-algebra and the uniform probability distribution). Let us now consider the random set $\Gamma' : [0, 1] \to \wp(\mathbb{R})$, defined as:

$$\Gamma'(\omega) = \begin{cases} \{0, k\} & \text{if } \omega \leq 2/3 \\ \{-k, 0, k\} & \text{if } \omega > 2/3 \end{cases}$$

We easily check that $\mathcal{P}(\Gamma') = \mathcal{M}(P^*_{\Gamma'}) = \mathcal{M}(P^*_{\Gamma}) \neq \mathcal{P}(\Gamma)$. We derive that the upper probability distributions of both random sets coincide ($P^*_{\Gamma'} = P^*_{\Gamma}$). Hence, they induce the same probability distribution on $\sigma_\wp$, i.e., $P \circ \Gamma^{-1} = P \circ \Gamma'^{-1}$. We see once more that, in general, the probability distribution induced by a random set does not determine its probability envelope.

3.5.3 Probability Envelopes Versus Credal Sets: Fuzzy Random Variables

Next we shall slightly modify the last example and consider a fuzzy random variable instead of a random set. We shall study the differences among $\pi_{\tilde{X}}, \pi^{\downarrow}_{\tilde{X}}$ and $\pi^*_{\tilde{X}}$ about the entropy of the original probability $H(P_{X_0})$. We shall compare the fuzzy sets $H(\pi_{\tilde{X}})$, $H(\pi^{\downarrow}_{\tilde{X}})$ and $H(\pi^*_{\tilde{X}})$, each one being of the form

$$H(\pi)(x) = \sup\{\pi(Q) : H(Q) = x\}, \ \forall x \in \mathbb{R}^+.$$

Example 6 Let Ω be the same universe and let $\Gamma : \Omega \to \wp(\mathbb{R})$ be the same random set as in Example 5. Let us define the fuzzy random variable $\tilde{X} : \Omega \to \mathcal{F}(\mathbb{R})$ as follows:

$$\tilde{X}(\omega_1)(x) = \begin{cases} 1 & \text{if } x = 0 \\ 0.5 & \text{if } x = k \\ 0 & \text{otherwise} \end{cases} \quad \tilde{X}(\omega_2)(x) = \begin{cases} 1 & \text{if } x = 0 \\ 0.5 & \text{if } x \in \{-k, k\} \\ 0 & \text{otherwise} \end{cases}$$

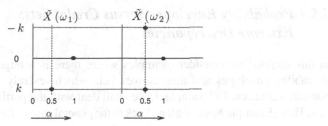

We observe that the α-cuts of $\tilde{X}$, $\tilde{X}_\alpha : \Omega \to \wp(\mathbb{R})$ are: $\tilde{X}_\alpha = \Gamma$, $\forall \alpha \in (0, 0.5]$ and $\tilde{X}_\alpha(\omega_i) = \{0\}$, $i = 1, 2$, $\forall \alpha \in (0.5, 1]$.

Hence, the classes of probability measures $\mathcal{P}(\tilde{X}_\alpha)$, $\mathcal{P}^\downarrow(\tilde{X}_\alpha)$ and $\mathcal{M}(P^*_{\tilde{X}_\alpha})$ are respectively equal to $\mathcal{P}(\Gamma)$, $\mathcal{P}^\downarrow(\Gamma)$, and $\mathcal{M}(P^*_\Gamma)$ in Example 5 if $\alpha \leq 0.5$, and $\mathcal{P}(\tilde{X}_\alpha) = \mathcal{P}^\downarrow(\tilde{X}_\alpha) = \mathcal{M}(P^*_{\tilde{X}_\alpha}) = \{(0, 1, 0)\}$ if $\alpha \in (0.5, 1)$.

So,

$$\mathcal{P}(\tilde{X}_\alpha) \subsetneq \mathcal{P}^\downarrow(\tilde{X}_\alpha) \subsetneq \mathcal{M}(P^*_{\tilde{X}_\alpha}) = \mathrm{Conv}(\mathcal{M}(P^*_{\tilde{X}_\alpha})), \; \forall \alpha \in (0, 0.5], \text{ and}$$

$$\mathcal{P}(\tilde{X}_\alpha) = \mathcal{P}^\downarrow(\tilde{X}_\alpha) = \mathcal{M}(P^*_{\tilde{X}_\alpha}), \; \forall \alpha \in (0.5, 1].$$

Let us now compare of $\pi_{\tilde{X}}$, $\pi^\downarrow_{\tilde{X}}$ and $\pi^*_{\tilde{X}}$. From the calculations of $\mathcal{P}(\tilde{X}_\alpha)$ and $\mathcal{P}^\downarrow(\tilde{X}_\alpha)$ for each α, we observe that $\pi^\downarrow_{\tilde{X}}$ strictly dominates $\pi_{\tilde{X}}$. In fact, $\pi^\downarrow_{\tilde{X}}((1/3, 1/3, 1/3)) = 0.5$ and $\pi_{\tilde{X}}((1/3, 1/3, 1/3)) = 0$. Furthermore, $\pi^*_{\tilde{X}}$ strictly dominates $\pi^\downarrow_{\tilde{X}}$. These differences are reflected in the calculus of the entropy. In fact, we easily check that $H(\pi_{\tilde{X}})(\log_2 3) = 0$.

But let us now consider the fuzzy set $H(\pi^\downarrow_{\tilde{X}})$: we observe that $H(\pi^\downarrow_{\tilde{X}})(\log_2 3) = 0.5$. We deduce that the second-order possibility measure $\pi^\downarrow_{\tilde{X}}$ provides less precise information about the original probability P_{X_0} than $\pi_{\tilde{X}}$ does. The fuzzy set $H(\pi^*_{\tilde{X}})$ is also more imprecise than $H(\pi_{\tilde{X}})$.

Remark 5 We can slightly modify the conditions in the last example and define another fuzzy random variable with the same probability distribution as $\tilde{X}$, but with a different fuzzy probability envelope, as we did in Remark 4 for the particular case of random sets. Let us consider the initial space $([0, 1], \beta_{[0,1]}, \lambda_{[0,1]})$ (the unit interval, with the usual Borel σ-algebra and the uniform probability distribution). Let us now consider the fuzzy random variable $\tilde{X}' : [0, 1] \to \mathcal{F}(\mathbb{R})$, defined as:

$$\tilde{X}'(\omega) = \begin{cases} \tilde{X}(\omega_1) & \text{if } \omega \leq 2/3 \\ \tilde{X}(\omega_2) & \text{if } \omega > 2/3 \end{cases}$$

We easily check that $\pi_{\tilde{X}'} = \pi^*_{\tilde{X}'} \neq \pi_{\tilde{X}}$. Furthermore, both fuzzy random variables induce the same probability distribution on $\sigma_{\mathcal{F}}$, i.e., $P \circ \tilde{X}^{-1} = P \circ \tilde{X}'^{-1}$. We conclude that, in general, the probability distribution induced by a fuzzy random variable does not determine its fuzzy probability envelope.

3.5.4 Probability Envelopes Versus Credal Sets: Extreme Discrepancies

In this section, we consider examples where there is a huge difference between probability envelopes and their convex hull, which severely impacts the range of possible variances. In Example 7, we shall deal with the particular case of random sets. Then (Example 8) we shall go back to the general case of fuzzy random variables.

Example 7 In this example, we shall consider two random sets with the same probability distribution (when regarded as classical measurable mappings). Nevertheless, their probability envelopes are totally different. Thus, if each of them is used to represent incomplete information about a random variable, the quality of the information provided by each one them is completely different from the other. Each random set will be defined on a specific universe.

(a) Let the first universe $\Omega_1 = \{a\}$ be a singleton. It represents a deterministic experiment. Let $\Gamma_1 : \Omega_1 \to \wp(\mathbb{R})$ be the multi-valued mapping defined as $\Gamma_1(a) = (-\infty, k]$, where k is an arbitrary positive number. Γ_1 represents the imprecise information about a certain constant $X_1(a) = x_1$.

(b) Let Ω_2 be the unit interval. Let us consider the usual Borel σ-algebra and the uniform distribution defined on it. Let $\Gamma_2 : \Omega_2 \to \wp(\mathbb{R})$ be the constant multi-valued mapping defined as $\Gamma_2(\omega) = (-\infty, k]$, $\forall \omega \in [0, 1]$. Γ_2 represents the imprecise information about a random variable, $X_2 : [0, 1] \to \mathbb{R}$. All we know about it is an upper bound (k) for its images $(X_2(\omega) \leq k, \ \forall \omega \in [0, 1])$.

Both multi-valued mappings are strongly measurable with respect to the respective initial σ-algebras and $\beta_{\mathbb{R}}$. Hence, it can be checked that they are measurable mappings for the σ-algebra generated by the class $\{\wp_A : A \in \beta_{\mathbb{R}}\}$, where $\wp_A = \{C \in \mathbb{R}^n : C \cap A \neq \emptyset\}$. They induce the same probability measure on this σ-algebra. In fact, both of them take the "value" $(-\infty, k]$ with probability 1. Hence, their upper probability measures, $P^*_{\Gamma_1}$ and $P^*_{\Gamma_2}$, coincide, and hence also $\mathcal{M}(P^*_{\Gamma_1}) = \mathcal{M}(P^*_{\Gamma_2})$ Nevertheless the classes $\mathcal{P}(\Gamma_1)$ and $\mathcal{P}(\Gamma_2)$ are not the same. We easily check that:

$$\mathcal{P}(\Gamma_1) = \{\delta_c : c \leq k\}.$$

(For each $c \in \mathbb{R}$, δ_c represents the probability distribution degenerated on c.) And, besides,

$$\mathcal{P}(\Gamma_2) = \{Q : Q([0, k]) = 1\}.$$

This reflects, for instance, that we know much more about the dispersion of X_1 than about the dispersion of X_2. In fact, we know that X_1 is a constant (it has zero variance). However, we have no information about the variance of X_2 (under the available information, it can be any non-negative number, hence arbitrary large).

Again we see that the probability distribution of a random set does not keep all the relevant information when it is regarded as the imprecise observation of a random variable.

Example 8 We can modify last example and consider two fuzzy random variables, instead of random sets. Let each one of them be defined on a different space:

(a) Let the first universe $\Omega_1 = \{a\}$ be a singleton. Let $\tilde{X}_1 : \Omega_1 \to \mathcal{F}(\mathbb{R})$ be the multi-valued mapping whose images are normal triangular fuzzy intervals defined as

$$\tilde{X}_1(a)(x) = \begin{cases} \frac{x}{k} & \text{if } x \in [0, k) \\ \frac{2k-x}{k} & \text{if } x \in [k, 2k], \end{cases}$$

where k is an arbitrary positive number. This fuzzy random variable represents the available imprecise information about some constant $X_1(a) = x_1$, hence the variance is known to be zero.

(b) Let Ω_2 be the unit interval. Let us consider the usual Borel σ-algebra and the uniform distribution defined on it. Let $\tilde{X}_2 : \Omega_2 \to \wp(\mathbb{R})$ be the constant fuzzy-valued mapping defined as $\tilde{X}_2(\omega) = \tilde{X}_1(a)$, $\forall \omega \in [0, 1]$. $\tilde{X}_2$ represents the imprecise information about a random variable, $X_2 : [0, 1] \to \mathbb{R}$.

Both of them take the "value" $\tilde{X}_1(a)$ with probability 1. (They induce the same probability measure on a σ-algebra defined on $\mathcal{F}(\mathbb{R})$.) Nevertheless the second-order possibility measures $\pi_{\tilde{X}_1}$ and $\pi_{\tilde{X}_2}$ are not the same. The information provided by $\pi_{\tilde{X}_1}$ is much more restrictive. In fact, it assigns zero possibility to each non-degenerate probability. Here, the information on the variance is imprecise. Its core is zero (this is the constant map $X(\omega) = k$), and its support is $[0, k^2]$.

3.5.5 Joint Probability Envelopes and Credal Sets of FRV's

Let us study the information about the joint probability measure associated to a pair of independent identically distributed random variables. We shall describe the information about each one of them by a random set (Example 9) or a fuzzy random variable (Example 10). We shall observe that the information provided by $\mathcal{M}(P_\Gamma^*)$ (resp. $\pi_{\tilde{X}}^*$) is again much more imprecise than the information contained in $\mathcal{P}(\Gamma)$ (resp. $\pi_{\tilde{X}}$.)

Example 9 Suppose an urn with 3 numbered balls. We know that all of them are coloured either red or white. Furthermore, we know that ball number 1 is red and ball number 3 is white. We do not know the colour of the second ball. We can represent our information about the colour of each ball by the multi-valued mapping $\Gamma : \{1, 2, 3\} \to \wp(\{r, w\})$ defined as

$$\Gamma(\{1\}) = \{r\}, \quad \Gamma(\{2\}) = \{r, w\}, \quad \Gamma(\{3\}) = \{w\}.$$

Let us identify each probability Q on $\wp(\{r, w\})$ by the pair $(Q(\{r\}), Q(\{w\}))$. The probability envelope associated to Γ is $\mathcal{P}(\Gamma) = \{(1/3, 2/3), (2/3, 1/3)\}$. This set of probability measures represents the available information about the experiment consisting in drawing a ball at random and observing its colour.

Let us now suppose that a ball is drawn at random from the urn and put back, and then a second ball is drawn at random (and the two drawings are independent). Let us identify now each probability measure Q on the product space by the tuple $(Q(\{rr\}), Q(\{rw\}), Q(\{wr\}), Q(\{ww\}))$. All we know about the true joint probability is that it belongs to the set

$$\{(p^2, p(1-p), (1-p)p, (1-p)^2) : (p, 1-p) \in \mathcal{P}(\Gamma)\}$$
$$= \{(1/9, 2/9, 2/9, 4/9), (4/9, 2/9, 2/9, 1/9)\}.$$

On the other hand, the set of probability measures on $\wp(\{r, w\})$ that are dominated by P_{Γ}^{*} is the convex hull of $\mathcal{P}(\Gamma)$,

$$\mathcal{M}(P_{\Gamma}^{*}) = \{(p, 1-p) : p \in [1/3, 2/3]\}.$$

Nevertheless, the class $\{(p^2, p(1-p), (1-p)p, (1-p)^2) : (p, 1-p) \in \mathcal{M}(P_{\Gamma}^{*})\}$ is not convex. In particular it is not included in the convex hull of $\{(p^2, p(1-p), (1-p)p, (1-p)^2) : (p, 1-p) \in \mathcal{P}(\Gamma)\}$.

In fact, $(0.25, 025, 0.25, 0.25)$ belongs to the first one, but not to the convex hull of the latter. Under the conditions of this experiment, we are sure that the probability of $\{rw\}$ is $Q(\{rw\}) = 2/9$. Nevertheless, when we consider the set $\mathcal{M}(P_{\Gamma}^{*})$ as the class of probability measures compatible with the experiment we lose some information about $Q(\{rw\})$: in that case, we can only say that it is between $2/9$ and 0.25. Hence, the information provided by the probability distribution induced by Γ (regarded as a $\wp(\{1, 2, 3\}) - \wp(\wp(\{r, w\}))$ measurable mapping), is not enough in this example.

Example 10 Let us now add some information to the last example. Assume, that, in addition to the information provided there, we now, with confidence 0.7 that the second ball is red. According to [14], the information about the experiment of drawing a ball at random can be now modeled by the fuzzy random variable: $\tilde{X} : \{1, 2, 3\} \to \Gamma(\{r, w\})$:

$$\tilde{X}(\{1\})(x) = \begin{cases} 1 & \text{if } x = r \\ 0 & \text{if } x = w \end{cases}$$

$$\tilde{X}(\{2\})(x) = \begin{cases} 1 & \text{if } x = r \\ 0.3 & \text{if } x = w \end{cases}$$

$$\tilde{X}(\{3\})(x) = \begin{cases} 1 & \text{if } x = w \\ 0 & \text{if } x = r \end{cases}$$

Let us first compute the second-order possibility distributions $\pi_{\tilde{X}}$ and $\pi_{\tilde{X}}^*$. Based on the calculations in the last example, they are defined as follows:

$$\pi_{\tilde{X}}((2/3, 1/3)) = 1, \pi_{\tilde{X}}((1/3, 2/3)) = 0 \text{ and } \pi_{\tilde{X}}((p, 1 - p)) = 0, \forall p \notin \{1/3, 2/3\}.$$

On the other hand,

$$\pi_{\tilde{X}}^*((p, 1 - p)) = 0.3, \ \forall p \in [1/3, 2/3), \text{ and } \pi_{\tilde{X}}^*((2/3, 1/3)) = 1.$$

Let us now assume, as in Example 9, that a ball is drawn at random from the urn and put back, and then a second ball is drawn again at random. The information about the true joint probability distribution is determined by the second-order possibility measure prod $\pi_{\tilde{X}}$ defined as

$$\text{prod } \pi_{\tilde{X}}((p^2, p(1 - p), (1 - p)p, (1 - p)^2)) = \pi_{\tilde{X}}((p, 1 - p)), \ \forall p \in [0, 1].$$

This information is quite different from the information described by

$$\text{prod } \pi_{\tilde{X}}^*((p^2, p(1 - p), (1 - p)p, (1 - p)^2)) = \pi_{\tilde{X}}^*((p, 1 - p)), \ \forall p \in [0, 1].$$

For instance,

$$\text{prod } \pi_{\tilde{X}}^*(0.25, 0.25, 0.25, 0.25) = 0.3, \text{ nevertheless}$$

$$\text{prod } \pi_{\tilde{X}}(0.25, 0.25, 0.25, 0.25) = 0.$$

3.5.6 Fuzzy Random Variables Versus Random Fuzzy Sets

In the last examples, we have compared the information provided by $\pi_{\tilde{X}}, \pi_{\tilde{X}}^{\downarrow}$ and $\pi_{\tilde{X}}^*$ about P_{X_0}. We have observed that, under the Kruse and Meyer interpretation, $\pi_{\tilde{X}}$ represents all the available information about this probability distribution. We have also noticed that, when we replace $\pi_{\tilde{X}}$ by $\pi_{\tilde{X}}^{\downarrow}$ or $\pi_{\tilde{X}}^*$, we can lose some relevant information. We have also shown that $\mathcal{M}(P_{\Gamma}^*)$ represents the same information as the probability measure that the random set induces on $\sigma_{\wp}$. Hence, the probability measure induced by the random set (considered as a measurable mapping) does not contain all the relevant information about the original probability, P_{X_0}. Returning to the general case of fuzzy random variables, we can also try to find the relationships between $\pi_{\tilde{X}}^*$ and the probability measure induced by $\tilde{X}$ on $\sigma_{\mathcal{F}}$, the algebra of measurable families of fuzzy subsets of U. We prove below that the probability distribution $P \circ \tilde{X}^{-1}$ determines $\pi_{\tilde{X}}^*$.

Proposition 3 *The second-order possibility distribution $\pi_{\tilde{X}}^*$ is determined by the induced probability distribution, $P \circ \tilde{X}^{-1}$ on fuzzy subsets of U.*

Proof Let us first remind that $\pi_{\tilde{X}}^*$ is given by the nested family of sets $\{\mathcal{M}(P_{\tilde{X}_\alpha}^*)\}_{\alpha \in [0,1]}$. In other words, it is determined by the family of upper probabilities $\{P_{\tilde{X}_\alpha}^*\}_{\alpha \in [0,1]}$. Furthermore, each one of these upper probabilities is, as well, determined by $P \circ \tilde{X}_\alpha^{-1}$. In fact:

$$P_{\tilde{X}_\alpha}^*(A) = P \circ \tilde{X}^{-1}(\mathcal{C}_\alpha^A). \qquad \qquad \square$$

But as we shall demonstrate, the converse is not true, the fuzzy credal set $\pi_{\tilde{X}}^*$ does not determine in general the induced probability distribution, $P \circ \tilde{X}^{-1}$. It does when, in particular, $\tilde{X}$ is a random set, but not in general.

We now show an example of two fuzzy random variables with the same convex fuzzy probability envelope, but inducing different probability measures on $\sigma_\mathcal{F}$. We define two fuzzy random variables, $\tilde{X}$ and $\tilde{Y}$ associated to the same family of upper probabilities, $\mathcal{M}(P_{\tilde{X}_\alpha}^*) = \mathcal{M}(P_{\tilde{Y}_\alpha}^*)$, $\forall \alpha$. Nevertheless, their induced probability distributions on fuzzy subsets of $\tilde{U}$ do not coincide.

Example 11 Let us consider a probability space, $(\Omega, \mathcal{A}, P)$ where $\Omega = \{\omega_1, \omega_2\}$, $\mathcal{A} = \wp(\Omega)$ and $P(\{\omega_1\}) = 0.5 = P(\{\omega_2\})$. Let us consider the fuzzy random variables $\tilde{X} : \Omega \to \mathcal{F}(\mathbb{R})$ and $\tilde{Y} : \Omega \to \mathcal{F}(\mathbb{R})$, whose α-cuts are defined as:

$$\tilde{X}_\alpha(\omega_1) = [0, 3], \alpha \le 0.5, \tilde{X}_\alpha(\omega_1) = [1, 2], \alpha > 0.5$$
$$\tilde{X}_\alpha(\omega_2) = [1, 3], \alpha \le 0.5, \tilde{X}_\alpha(\omega_2) = [2, 3], \alpha > 0.5$$
$$\tilde{Y}_\alpha(\omega_1) = [0, 3], \alpha \le 0.5, \tilde{Y}_\alpha(\omega_1) = [2, 3], \alpha > 0.5$$
$$\tilde{Y}_\alpha(\omega_2) = [1, 3], \alpha \le 0.5, \tilde{Y}_\alpha(\omega_2) = [1, 2], \alpha > 0.5$$

We observe that they induce different probability distributions on $(\mathcal{F}(\mathbb{R}), \sigma_\mathcal{F})$. On the other hand their associated second-order possibility distributions coincide. Indeed note that the (fuzzy) outcomes of $\tilde{X}$ are different from those of $\tilde{Y}$, and therefore both probability distributions are different; equivalently the two fuzzy belief functions have different fuzzy focal sets pictured on Fig. 3.2. But, we can check that the following equalities hold:

$$P(\omega_1) = P(\omega_2) = 0.5$$
$$\tilde{X}_\alpha(\omega_1) = \tilde{Y}_\alpha(\omega_1), \ \tilde{X}_\alpha(\omega_2) = \tilde{Y}_\alpha(\omega_2), \ \forall \alpha \le 0.5$$
$$\tilde{X}_\alpha(\omega_1) = \tilde{Y}_\alpha(\omega_2), \ \tilde{X}_\alpha(\omega_2) = \tilde{Y}_\alpha(\omega_1), \ \forall \alpha > 0.5.$$

In other words, the random sets induced at levels $\alpha = 0.5$ and $\alpha = 1$ by $\tilde{X}$ and $\tilde{Y}$ have the same behaviour. So the corresponding probability sets with support in $[0, 3]$

Fig. 3.2 Two fuzzy random variables with different probability distributions associated to the same nested family of envelopes

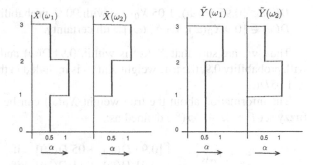

are also the same that is $\mathcal{P}(\tilde{X}_\alpha) = \mathcal{P}(\tilde{Y}_\alpha)$, $\forall \alpha$, which shows that the two higher order possibility distributions are also the same.

So, the second-order possibility distributions associated to a fuzzy random variable, $\tilde{X}$, do not determine its induced probability measure ($P \circ \tilde{X}^{-1}$) even if the latter determines the second-order possibility distribution $\pi_{\tilde{X}}^*$. But, as we already know from Remark 5 and Example 8, $P \circ \tilde{X}^{-1}$ does not determine the second-order possibility distributions $\pi_{\tilde{X}}$ and $\pi_{\tilde{X}}^{\downarrow}$.

3.5.7 A Summary Example

In the last examples we have shown the differences between the information provided by the fuzzy probability envelope and the classical probability distribution induced by a fuzzy random variable. Next, we shall illustrate the main ideas in the chapter with an additional example. We shall show the usefulness of the fuzzy probability envelope as the representation of the available imprecise information about the true probability distribution in a random experiment. We shall represent the imprecise information about the weight of certain objects chosen at random by means of a fuzzy random variable. First we shall show how the fuzzy random variable is derived from the imprecise data. We shall then calculate the associated second-order possibility distributions, as well as the induced probability measure on $(\mathcal{F}(\mathbb{R}), \sigma_{\mathcal{F}})$. We shall compare the information that these different models provide about the true probability distribution associated to the random experiment.

Example 12 Let us consider a population of objects, Ω, and we get imprecise information on their weights. Let us denote by $X_0(\omega)$ the (true) weight of an arbitrary object $\omega \in \Omega$. However, we measure this weight with scales of low precision, so that an expert tells us that 90 % of the measurements are in a 5 % error margin of the observed value $D(\omega)$. But in general, we can always guarantee an error less than 10 %. We do not assume the error is a matter of noise (it is for instance an ill-known bias). So we know that

$D(\omega) \in [0.95\, X_0(\omega), 1.05\, X_0(\omega)]$ with 90 % probability, and
$D(\omega) \in [0.9\, X_0(\omega), 1.1\, X_0(\omega)]$ with certainty.

Thus, we are sure that $X_0(\omega)$ is within $0.91\, D(\omega)$ and $1.11\, D(\omega)$. Furthermore, with probability 0.9, the true weight $X_0(\omega)$ is included in the interval from $0.95\, D(\omega)$ to $1.05\, D(\omega)$.

The information about the true weight $X_0(\omega)$ can be described by the random fuzzy set $\tilde{\Gamma} : \Omega \to \wp(\mathbb{R})$ defined as:

$$\tilde{\Gamma}_\alpha(\omega) = \begin{cases} [0.95\, D(\omega), 1.05\, D(\omega)] & \text{if} \quad \alpha \le 0.9 \\ [0.91\, D(\omega), 1.11\, D(\omega)] & \text{if} \quad \alpha > 0.9, \end{cases}$$

This random fuzzy set reflects the following information: In 90 % of the time, the true weight is within $0.95\, D(\omega)$ and $1.05\, D(\omega)$. In the remaining 10% of the time, we can only be sure that it is within $0.91\, D(\omega)$ and $1.11\, D(\omega)$.

Summarizing, the information about the random variable $X_0 : \Omega \to \mathbb{R}$ is described by the fuzzy random variable $\tilde{X} : \Omega \to \mathcal{F}(\mathbb{R})$, where $\tilde{X}(\omega)$ is a possibility distribution of weights for each $\omega \in \Omega$ defined as

$$\tilde{X}(\omega)(x) = \begin{cases} 0 & \text{if} \quad x \notin [0.91\, D(\omega), 1.11\, D(\omega)] \\ 0.1 & \text{if} \quad x \in [0.91\, D(\omega), 1.11\, D(\omega)] \setminus [0.95\, D(\omega), 1.05\, D(\omega)]. \\ 1 & \text{if} \quad x \in [0.95\, D(\omega), 1.05\, D(\omega)] \end{cases}$$

For each pair (ω, x), $\tilde{X}(\omega)(x)$ represents the grade of possibility that the true weight of the object ω, $X_0(\omega)$, coincides with x. In other words, for a confidence level $1 - \alpha$ all we now about X_0 is that it belongs to the set

$$S(\tilde{X}_\alpha) = \{X : \Omega \to \mathbb{R} \text{ such that} : X(\omega) \in [\tilde{X}(\omega)]_\alpha, \, \forall \omega \in \Omega\}.$$

Once we have represented the available information about X_0 by means of a fuzzy random variable, let us describe the (imprecise) available information about its induced probability measure, P_{X_0}. Let us suppose that Ω contains only 10 objects, $\omega_1, \dots, \omega_{10}$ and the respective values displayed by the scale have been

500, 300, 400, 500, 500, 400, 400, 300, 300, 400.

Hence, the random variable D takes the values $d_1 = 300$, $d_2 = 400$, and $d_3 = 500$ with respective probabilities $0.3, 0.4$, and 0.3. Thus, the fuzzy random variable $\tilde{X}$ also takes three different "values" in $\mathcal{F}(\mathbb{R})$, π_1, π_2 and π_3. Their membership functions are, respectively:

$$\pi_1(x) = \begin{cases} 0 & \text{if } x \notin [0.91\, d_1, 1.11\, d_1] \\ 0.1 & \text{if } x \in [0.91\, d_1, 1.11\, d_1] \setminus [0.95\, d_1, 1.05\, d_1] \\ 1 & \text{if } x \in [0.95\, d_1, 1.05\, d_1] \end{cases}$$

$$\pi_2(x) = \begin{cases} 0 & \text{if } x \notin [0.91\,d_2, 1.11\,d_2] \\ 0.1 & \text{if } x \in [0.91\,d_2, 1.11\,d_2] \setminus [0.95\,d_2, 1.05\,d_2] \\ 1 & \text{if } x \in [0.95\,d_2, 1.05\,d_2] \end{cases}$$

$$\pi_3(x) = \begin{cases} 0 & \text{if } x \notin [0.91\,d_3, 1.11\,d_3] \\ 0.1 & \text{if } x \in [0.91\,d_3, 1.11\,d_3] \setminus [0.95\,d_3, 1.05\,d_3] \\ 1 & \text{if } x \in [0.95\,d_3, 1.05\,d_3] \end{cases}$$

$\tilde{X}$ takes these "values" with respective probabilities $p_1 = 0.3$, $p_2 = 0.4$, and $p_3 = 0.3$. The mass function $p(\pi_i) = p_i$, $i = 1, 2, 3$ determines the probability distribution on $\sigma_{\mathcal{F}}$ associated to $\tilde{X}$, $P \circ \tilde{X}^{-1}$. But this probability measure does not describe the information that $\tilde{X}$ captures about P_{X_0}. In fact, and according to the information given above, for each confidence level $1 - \alpha$, all we now about P_{X_0} is that it belongs to the set of probability functions:

$$\mathcal{P}(\tilde{X}_\alpha) = \{P_X : X \in S(\tilde{X}_\alpha)\}.$$

Hence, the available information about P_{X_0} should be described by means of the second-order possibility distribution $\pi_{\tilde{X}}$ (the fuzzy set associated to the graded representation $\{\mathcal{P}(\tilde{X}_\alpha)\}_{\alpha \in [0,1]}$. Let us describe $\mathcal{P}(\tilde{X}_\alpha)$ for each $\alpha \in [0, 1]$).

- For $\alpha \le 0.1$, $Q \in \mathcal{P}(\tilde{X}_\alpha)$ if and only if:

  $\exists\, x_1, x_2, x_3 \in [0.91\,d_1, 1.11\,d_1] : \sum_{i=1}^{3} Q(\{x_i\}) = 0.3$

  $\exists\, x_4, x_5, x_6, x_7 \in [0.91\,d_2, 1.11\,d_2] : \sum_{i=4}^{7} Q(\{x_i\}) = 0.4$

  $\exists\, x_8, x_9, x_{10} \in [0.91\,d_3, 1.11\,d_3] : \sum_{i=8}^{10} Q(\{x_i\}) = 0.3.$

- For $\alpha > 0.1$, $Q \in \mathcal{P}(\tilde{X}_\alpha)$ if and only if:

  $\exists\, x_1, x_2, x_3 \in [0.95\,d_1, 1.05\,d_1] : \sum_{i=1}^{3} Q(\{x_i\}) = 0.3$

  $\exists\, x_4, x_5, x_6, x_7 \in [0.95\,d_2, 1.05\,d_2] : \sum_{i=4}^{7} Q(\{x_i\}) = 0.4$

  $\exists\, x_8, x_9, x_{10} \in [0.95\,d_3, 1.05\,d_3] : \sum_{i=8}^{10} Q(\{x_i\}) = 0.3.$

In other words, $\pi_{\tilde{X}}$ takes the values 0, 0.1 and 1. These values are assigned as follows:

$$\pi_{\tilde{X}}(Q) = \begin{cases} 0 & \text{if } Q \notin \mathcal{P}(\tilde{X}_{0.1}) \\ 0.1 & \text{if } Q \in \mathcal{P}(\tilde{X}_{0.1}) \setminus \mathcal{P}(\tilde{X}_1) \\ 1 & \text{if } Q \in \mathcal{P}(\tilde{X}_1) \end{cases}$$

It is easy to check that $\pi_{\overset{\downarrow}{\tilde{X}}}$ coincides with $\pi_{\tilde{X}}$ in this example. But $\pi^*_{\tilde{X}}$ strictly contains them. We can observe that:

$$\pi^*_{\tilde{X}}(Q) = \begin{cases} 1 & \text{if } Q(A_{11}) = 0.3, \ Q(A_{21}) = 0.4 \text{ and } Q(A_{31}) = 0.3 \\ 0.1 & \text{if } Q(A_{12}) = 0.3, \ Q(A_{22}) = 0.4 \text{ and } Q(A_{32}) = 0.3, \text{ but } \pi^*_{\tilde{X}}(Q) \neq 1 \\ 0 & \text{otherwise.} \end{cases}$$

where:

$$A_{11} = [0.95d_1, 1.05d_1] \quad A_{21} = [0.95d_2, 1.05d_2] \quad A_{31} = [0.95d_3, 1.05d_3]$$
$$A_{12} = [0.91d_1, 1.11d_1] \quad A_{22} = [0.91d_2, 1.11d_2] \quad A_{32} = [0.91d_3, 1.11d_3].$$

Notice that the information provided by $\pi_{\tilde{X}}$ is more precise than the information provided by $\pi^*_{\tilde{X}}$. In fact, $\pi^*_{\tilde{X}}$ does not distinguish whether the original probability distribution is discrete or continuous. The same happens with the probability distribution induced by $\tilde{X}$, $P \circ \tilde{X}^{-1}$. It does not take into account the nature of the initial space. The same probability distribution could be generated from a non-atomic initial space. In that case, the original probability measure could be continuous, and $\pi_{\tilde{X}}$ would be convex, i.e.,

$$\pi_{\tilde{X}}(\lambda\, Q_1 + (1-\lambda)\, Q_2) \geq \lambda\pi_{\tilde{X}}(Q_1) + (1-\lambda)\, Q_2, \ \forall \lambda \in [0,1].$$

However, in our example, $\pi_{\tilde{X}}$ is not convex, because the original probability is focused on no more than 10 points.

3.6 Fuzzy Random Variables as Crisp Credal Sets

So far in this book, the set-up for the representation of epistemic random sets and fuzzy random variables $\tilde{X}$ was assuming that variability was due to the sample space only, while the higher order uncertainty reflected the ill-perception of the otherwise deterministic value of $X(\omega)$, once ω is fixed. However, while remaining in accordance with the possibilistic interpretation of fuzzy sets, the information provided by a fuzzy random variable $\tilde{X}$ could be interpreted in a slightly different way following the path initiated in [7, 13]. Suppose that the displayed quantity $X(\omega)$ for a specific ω is itself a random variable (affected by other factors, for instance a noisy measurement process). Now, there is a sequence of two (independent) random phenomena whose sample spaces are Ω and U, respectively. But while the probability distribution on $(\Omega, \mathcal{A})$, $P : \mathcal{A} \to [0, 1]$, is completely determined, the noisy measurement process is only known via a family of conditional possibility measures $\{\Pi(\cdot|\omega)\}_{\omega\in\Omega}$, each of them being encoded by the fuzzy set $\tilde{X}(\omega)$. This is another understanding of the fuzzy mapping relating Ω and U: It restricts, for each value ω in the sample set Ω, a family of conditional probability measures $P(\cdot \mid \omega)$ dominated by $\Pi(\cdot \mid \omega)$.

Each conditional possibility measure $\Pi(\cdot \mid \omega)$ models our knowledge about a noisy measurement outcome on U. If the result of the first experiment is ω, then the possibility degree of x occurring in the second one is $\tilde{X}(\omega)(x) = \Pi(x \mid \omega)$. In other words, we know the probability measure that drives the primary random process, but the measurement process of outcomes is tainted with uncertainty. More specifically:

- The marginal probability $P : \mathcal{A} \to [0, 1]$ is completely known.
- There exists a *transition probability* that models the relationship between the outcomes of both sub-experiments, $P_U : \mathcal{A}' \times \Omega \to [0, 1]$, i.e.:

  - $P_U(\cdot \mid \omega)$ is a probability measure, $\forall \omega \in \Omega$.
  - $P_U(B \mid \cdot)$ is $\mathcal{A} - \beta_{[0,1]}$ measurable.
  - Our imprecise knowledge about P_U is determined by the following inequalities:

$$P_U(B|\omega) \le \Pi(B|\omega) = \sup_{b \in B} \tilde{X}(\omega)(b), \ \forall \omega \in \Omega, \ \forall B \in \mathcal{A}'.$$

Remark 6 In the other sections of the book, $\forall \omega \in \Omega$, the membership function of $\tilde{X}(\omega)$ is viewed as a possibility distribution that restricts the unknown *deterministic* value $X_0(\omega)$ (a Dirac conditional probability $P(\{X_0(\omega)\}|\omega) = 1$). The fuzzy set $\tilde{X}(\omega)$ represents, in that case, the second-order conditional possibility that assigns to every Dirac distribution δ_x the possibility degree $\tilde{X}(\omega)(x)$. Or equivalently, the second-order conditional necessity that assigns a necessity degree greater than or equal to $1 - \alpha$ to each level set of Dirac distributions $\{\delta_x : x \in \tilde{X}_\alpha(\omega)\}$. Here, we understand $\tilde{X}(\omega)$ as the (first-order) conditional possibility distribution $\Pi(\cdot \mid \omega)$ restricting a conditional probability $P_U(\cdot|\omega)$, that, for instance, describes a measurement process that is not assumed to be deterministic. In other words, the conditional possibility distribution is understood as a crisp convex set of objective probability measures.

As a consequence, we have partial information about the probability distribution P_U on $(U, \mathcal{A}')$, resulting from the sequence of two random experiments. The combination of information coming from both sources, using natural extension techniques [73], allows to describe the available information about this probability distribution by means of an upper probability. In fact, this probability measure is given by the formula:

$$P_U(B) = \int_\Omega P_U(B|\omega) \, dP(\omega),$$

where

$$P_U(B|\omega) \le \Pi(B|\omega) = \sup_{b \in B} \tilde{X}(\omega)(b), \ \forall \omega \in \Omega, \ \forall B \in \mathcal{A}'.$$

Thus, the upper and lower bounds for the probability of an arbitrary event $B \in \mathcal{A}'$ are:

$$\overline{P}(B) = \sup_{P_U \in \mathcal{H}} \int_{\Omega} P_U(B, \omega) \, dP(\omega); \quad \underline{P}(B) = \inf_{P_U \in \mathcal{H}} \int_{\Omega} P_U(B, \omega) \, dP(\omega) \quad (3.5)$$

where the set of transition probabilities is:

$$\mathcal{H} = \{P_U : P_U(B|\omega) \leq \Pi(B|\omega), \; \forall \omega \in \Omega, \; \forall B \in \mathcal{A}'\}. \quad (3.6)$$

A generic example of such a situation is when the outcome X on U depends not only on a random element $\omega \in \Omega$ with known distribution, but also on some other random parameter $\theta \in \Theta$ the value of which is partially known as restricted by a fuzzy set F. So $X = f(\omega, \theta)$ for some function f. Then the conditional possibility distribution is simply defined by the extension principle $\pi(r|\omega) = \sup_{\theta: \, r=f(u,\omega)} \mu_F(\theta)$. The underlying conditional probability distribution is $P_U(B|\omega) = \int_B dP_U(\{u : r = f(\theta, \omega)\})$.

Under fairly general conditions (see [13] for a detailed analysis) the following equalities hold:

$$\overline{P}(B) = \int_{\Omega} \Pi(B|\omega) \, dP(\omega) \quad \text{and} \quad \underline{P}(B) = \int_{\Omega} N(B|\omega) \, dP(\omega) \quad (3.7)$$

Let us assume that, in particular, the FRV takes a finite number of different imprecise values with membership functions $\mu_1, \ldots, \mu_n$ having respective probabilities $p_1, \ldots, p_n$. Namely, the initial universe Ω is partitioned into n subsets $A_1, \ldots, A_n$ such that $P(A_i) = p_i$ and $\tilde{X}(\omega)$ is the fuzzy set with membership function μ_i, for each $\omega \in A_i$. Then, according to Eq. (3.7) the upper and lower probabilities are determined as follows [6]:

$$\overline{P}(B) = \sum_{i=1}^{n} p_i \sup_{x \in B} \mu_i(x) \quad \text{and} \quad \underline{P}(B) = \sum_{i=1}^{n} p_i \inf_{x \in B} \mu_i(x) \quad (3.8)$$

In fact, when the result of the first sub-experiment belongs to the set A_i, the probability of occurrence of B is less than or equal to $\Pi_i(B) = \sup_{x \in B} \mu_i(x)$. For each $i \in \{1, \ldots, n\}$ the probability of getting an element in A_i is $P(A_i) = p_i$. Hence the true probability of B is less or equal to $\sum_{i=1}^{n} p_i \sup_{x \in B} \mu_i(x)$. According to Eqs. (3.7) and (3.8), this quantity appears to be the tightest upper bound for the probability of occurrence of B in the second sub-experiment. A similar conclusion can be drawn for the lower bound. Note that we get expressions that are those described in Remark 3, which shows that $\overline{P}(B)$ and $\underline{P}(B)$ have the same properties as plausibility and belief functions respectively.

The above formulae can be generalized for infinite universes. In [16] we prove that the upper probability $\overline{P}(B)$ can be written, in general, as the Lebesgue integral:

$$\overline{P}(B) = \int\limits_0^1 P^*_{\tilde{X}_\alpha}(B)\, d\lambda(\alpha),$$

where $P^*_{\tilde{X}_\alpha}$ represents the upper probability (in Dempster's sense) associated to the multi-valued mapping $\tilde{X}_\alpha$, i.e.

$$P^*_{\tilde{X}_\alpha}(B) = P(\{\omega \in \Omega : \tilde{X}_\alpha(\omega) \cap B \neq \emptyset\}).$$

Furthermore, $\overline{P}(B)$ coincides with the (Dempster's) upper probability associated to the multi-valued mapping $\Gamma : \Omega \times [0, 1] \to \wp(\mathbb{R})$ defined as follows:

$$\Gamma(\omega, \alpha) = \tilde{X}_\alpha(\omega) = \{u \in U : \tilde{X}(\omega) \geq \alpha\}. \tag{3.9}$$

Let us fix an arbitrary $\omega \in \Omega$. For each α, the probability that the image of $\Gamma(\omega, \cdot)$ is included in $\tilde{X}_\alpha(\omega)$ is greater or equal to $1 - \alpha$. Thus, Γ reflects our information about the outcome in the second sub-experiment, in accordance with the interpretation of FRV's presented in this subsection. When we regard the FRV as a family of conditional probability measures, it represents the same information as the multi-valued mapping Γ defined in Eq. (3.9).

Remark 7 If we use the set-up of this section with crisp imprecise information about the noisy measurement process described by the multivalued mapping Γ, the set of probabilities generated on U is convex and it is precisely the credal set of Γ defined in the previous chapter, that is the convex closure of the probability set induced by the selections of Γ in the finite setting or when the realisations of the random set are compact subsets of $\mathbb{R}^n$.

Example 13 Let us consider a variant of the previous example of measurement of the weight of objects. But now we assume that the measurement process is noisy. If the real weight of ω is $X_0(\omega) = x$, then the scale delivers a measurement $D(x)$ that follows a certain probability distribution over possible weights in $[0, \infty)$, that depends on the real weight x. However this probability distribution is ill-known and the device guidebook announces that 90 % of the measurements are in a 5 % error margin. But in general, it guarantees a random error less than 10 %. If the conditional probability of D has conditional density $p(y|x)$, then the probability distribution on the weight of ω would have density $p(y|X_0(\omega))$. The random process combining the picking of an object in Ω and the measurement of its weight would yield probability measure Q with density $q(y) = \int_\Omega p(y|X_0(\omega))\, dP(\omega)$ for the random variable $D(X_0)$ describing the measurement result.

Actually the information about $P(B|x)$ is that it is described by the conditional possibility measure $\Pi(B|x)$ with possibility distribution $\pi(y|x) = 1$ if $y \in [0.95x, 1.05x]$, 0 if $y \notin [0.90x, 1.10x]$, 0.1 otherwise. The information about the measurement can thus also be described by the set of conditional probabilities:

$$\{P(\cdot|x) : P([0, 95x, 1.05x]|x]) \geq 0.95 \text{ and } P([0.91x, 1.11x]|x) = 1\}.$$

The information about the weight of objects X_0 is thus represented by the credal set

$$\{Q(\cdot) = \int_{\Omega} P(\cdot|X_0(\omega)) \, dP(\omega) : P([0, 95x, 1.05x|x]) \geq 0.95 \text{ and } P([0.91x, 1.11x]|x) = 1\}.$$

3.7 Conclusion

This chapter has presented an epistemic point of view on fuzzy random variables in the tradition of Kwakernaak and Kruse, in contrast with the current mainstream literature on the topic [44, 72], that follows the random set tradition and the works of Puri and Ralescu. This book has tried to show that the confusion between these two traditions may have important practical consequences when modeling real situations. We proposed a collection of small examples to convince the reader that a set-valued variable expressing knowledge about a precise underlying one should not be handled as a precise random variable representing an attribute which is naturally set-valued. In the last case, fuzzy data can be handled by classical methods like any other kind of complex random data, while in the case of imprecise data, the classical statistical analysis must be carried out along with sensitivity analysis, as one question of interest can be to evaluate the impact of the incomplete knowledge on the results we could have obtained, had the data been precise. This approach is in the tradition of Dempster's upper and lower probabilities.

But the ontic versus epistemic distinction made in this book is not exactly the end of the story. Ontic random sets may not just model sets that occur as such in the nature. There are circumstances when epistemic set-valued data, even if they are imprecise descriptions of otherwise point-valued variables can be treated as ontic entities. Here are two examples:

- Suppose one imprecisely measures a precisely defined attribute a number of times, say via a number of different observers (e.g. human testimonies on the value of a quantity), but the actual aim of the statistics is to model the variability in the imprecision of the observers. In other words, while such fuzzy data are subjective descriptions of an otherwise objective quantity, they can be considered as ontic with respect to the observers. In particular, if agreeing observers provide nested set-valued estimates of a constant but ill-known value (say, each with full confidence), one may consider that the various levels of precision correspond to a form of variability, which justifies the use of a scalar distance between such sets in the computation of the variance. But it is the variance of the imprecision levels of the observer responses that is obtained, not the one pertaining to the variability of the objective underlying quantity.
- Sometimes human perceptions refer to a complex matter that cannot be naturally represented by a precise numerical value. For instance, ratings in a dish tasting

experiment are verbal rather than numerical. Imprecise terms then refer to no clear objective feature of the phenomenon under study. For instance, the taste of a dessert does not directly describe the objective ingredients of the dish. So, the collected imprecise data can be considered ontic, because you want to know if people will like the dessert, not how much butter or sugar it contains. Here again, the human perceptions can be handled as ontic entities. However, the question is then to figure out whether the collected subjective data in this kind of situation is liable of a representation by means of a fuzzy set over a numerical scale: indeed, the very reason why a precise numerical estimate is inappropriate in this kind of situation is because a one-dimensional numerical scale does not make sense. Then, the statistician is not better off when representing human perceptions by means of fuzzy sets (let alone trapezoidal ones) on a meaningless numerical scale.

• Moral [60] suggests that the distinction between ontic and epistemic sets can be exemplified by the two usual views of sets of probability functions, namely the original one by Dempster whereby the upper and lower probabilities bracket a real (objective) probability measure describing some random phenomenon, and the Shafer and Walley views whereby the lower probability represents a degree of belief without assuming the existence of an underlying real probability function. The credal set of probabilities that dominate the belief function is epistemic in the first case (as only one of the probability function is correct), and is ontic in the second case, since none of the probability functions inside the credal set is enough to represent the agent's epistemic state. The credal set is just a mathematical characterisation of the state of knowledge of the agent, more precisely an encoding of the set of gambles desirable for the agent.

These examples highlight the fact that the distinction ontic/epistemic does not match the distinction objective/subjective. In summary, a set-valued statistic is ontic if the set representation captures the essence of the issue under study; it is epistemic if its purpose is to provide some information on some precise entity that could not be precisely observed because of the poor quality of the knowledge. Adopting an ontic approach to the statistics of human perceptions of otherwise objective quantities yields a description of the observer behaviour, not of the natural phenomenon on which this observer reports.

Among potential applications of fuzzy random variables are statistical processing techniques such as regression, where again the two views (ontic vs. epistemic) are at odds. Many works follow the tradition of fuzzy least squares of Diamond [23, 26], that relies on the use of a precise variance of interval or fuzzy set-valued data, and recommends minimizing a scalar mean squared error [41–43]. This is a direct adaptation of the classical regression methodology to set-valued data, that tries to fit a fuzzy set-valued linear function on the fuzzy data. This methodology has been applied to other kind of data interpolation methods such as kriging [24, 25]. Fuzzy regression under the epistemic view is standard regression augmented with sensitivity analysis, so as to picture all the regression lines that would have been obtained, had the data been precisely measured. This approach has not been very much popular in the fuzzy regression literature, but it has been put to work in kriging methods for spatial interpolation [3–5, 55]. The main objection to such a sensitivity approach is the risk of getting very uninformative, hence useless results. However, it has been pointed out [47] that in the case of regression, and more generally learning methods, the model assumption (e.g. the shape of the distribution, linearity, etc.) should be instrumental in disambiguating fuzzy data, thus improving precision.

3.8 Exercises

1. Fuzzy random variables admit several interpretations (see [11]), depending on the problem we want to solve. Consider the following interpretations:

   • Ontic random fuzzy set (random object, Puri and Ralescu [62], Colubi et al. [10]).
   • Ill-known random variable (Kruse and Meyer [53]; Baudrit et al. [7]; Couso et al. [15]; Couso and Sánchez [16]).

   Choose the most appropriate of the above interpretations in the following examples:

   (a) The population is a set of people. The final set is a set of languages. The fuzzy random variable $\tilde{X}$ assigns, to each person, a fuzzy subset of the set of languages. Therefore, for each person, a membership value is assigned to each language. Such membership value represents a degree of preference for the language in a [0, 1] scale, related to her/his skill's degree using that language. Thus, $\tilde{X}(\omega)(l) > \tilde{X}(\omega)(l')$ means that ω (say, for instance, John) prefers to speak the language l than the language l', because he is more familiar with it. Those degrees of preference can be determined as a function of the CEFR levels, for instance.
   (b) A vehicle moves from one point to another. A GPS detects its position with some imprecision. In fact, it attaches a number of confidence degrees to a family of nested circles. The information provided by the GPS about a specific position can be represented by means of a fuzzy set (understood as a possibility measure). The population is a set of instants of time, and the fuzzy random variable assigns to each the fuzzy set representing the imprecise information provided by the GPS about the vehicle position.

2. Consider a GPS that provides the location of a point p with some imprecision. For the sake of simplicity, let us assume a one dimension problem ($p \in \mathbb{R}$). The GPS displays the point $d_0 \in \mathbb{R}$. According to the technical specifications, our information about the actual location, p, can be summarized as follows:

$$P([d_0-5, d_0+5]) = 1, \; P([d_0-3, d_0+3]) \geq 0.95, \; P([d_0-2, d_0+2]) \geq 0.65.$$

(a) Determine the possibility distribution associated to this class of probability measures (the most specific possibility dominating all those probability measures).

Hint: Take into account that the above conditions imply:

- $P((-\infty, d_0 - 5) \cup (d_0 + 5, \infty)) = 0$,
- $P((-\infty, d_0 - 3) \cup (d_0 + 3, \infty)) \leq 0.05$, and
- $P((-\infty, d_0 - 2) \cup (d_0 + 2, \infty)) \leq 0.35$.

(b) Calculate the degree of possibility of the interval $[d_0 - 5, d_0 - 2.5]$.

3. Check that the probability distribution induced by a trapezoidal fuzzy random variable (a fuzzy random variable whose outcomes are trapezoidal fuzzy subsets of the real line) is determined by the joint distribution of a 4-dimensional random vector. Derive, as a particular case, the fact that the probability distribution induced by an arbitrary triangular fuzzy random variable is univocally determined by the distribution of a 3-dimensional random vector.

4. Consider the information provided in Example 12 about the weight of 10 objects. Explain, with words, what kind of information would represent the scalar variance and the fuzzy-valued variance (based on Kruse's definition) in this case.

5. Check that two triangular fuzzy random variables are independent if and only the pair of 3-dimensional random vectors determined by their respective "extremes" and "modes" are stochastically independent.

References

1. J. Aumann, Integral of set valued functions. J. Math. Anal. Appl. **12**, 1–12 (1965)
2. J.-P. Auray, H. Prade. Robert Féron: a pioneer in soft methods for probability and statistics. In D. Dubois et al. (eds.), *Soft Methods for Handling Variability and Imprecision* (Proc. SMPS 2008), Advances in Soft Computing, 48, pp. 27–32. Springer (2008)
3. A. Bardossy, I. Bogardi, W.E. Kelly, Imprecise fuzzy information in geostatistics. Math. Geol. **20**, 287–311 (1988)
4. A. Bardossy, I. Bogardi, W.E. Kelly, Kriging with imprecise (fuzzy) variograms. I: theory. Math. Geol. **22**, 63–79 (1990)
5. A. Bardossy, I. Bogardi, W.E. Kelly, Kriging with imprecise (fuzzy) variograms. II: application. Math. Geol. **22**, 81–94 (1990)
6. C. Baudrit, D. Dubois, D. Guyonnet, H. Fargier. Joint treatment of imprecision and randomness in uncertainty propagation. In: Modern Information Processing: From Theory to Applications. B. Bouchon-Meunier, G. Coletti, R.R. Yager (Eds.), Elsevier, 37–47 (2006)
7. C. Baudrit, I. Couso, D. Dubois, Joint propagation of probability and possibility in risk analysis: towards a formal framework. Int. J. Approximate Reasoning **45**, 82–105 (2007)
8. L. Boyen, G. de Cooman, E. E. Kerre. On the extension of P-consistent mappings. In: G. de Cooman, D. Ruan, E. E. Kerre (Eds.), *Foundations and Applications of Possibility Theory-Proceedings of FAPT'95*, World Scientific (Singapore, 1995) 88–98.
9. J. Casillas L. Sánchez. Knowledge extraction from fuzzy data for estimating consumer behavior models. In: Proceedings of 2006 IEEE International Conference on Fuzzy Systems (FUZZ-IEEE'06). Vancouver, Canada, pp. 572–578 (2006)
10. A. Colubi, R. Coppi, P. D'Urso, M.A. Gil. Statistics with fuzzy random variables, METRON–Int. J. Stat. vol. LXV, 277–303 (2007)

11. I. Couso, D. Dubois, On the variability of the concept of variance for fuzzy random variables. IEEE Trans. Fuzzy Syst. **17**, 1070–1080 (2009)

12. I. Couso, D. Dubois, S. Montes, L. Sánchez. On various definitions of the variance of a fuzzy random variable, in: Proceedings of Fifth International Symposium on Imprecise Probabilities, Theory and Applications (ISIPTA 07) Prague, Czech Republic, pp. 135–144 (2007)

13. I. Couso, E. Miranda, G. de Cooman, A possibilistic interpretation of the expectation of a fuzzy random variable, in *Soft Methodology and Random Information Systems*, ed. by M. López-Daz, M.A. Gil, P. Grzegorzewski, O. Hryniewicz, J. Lawry (Springer, Berlin, 2004), pp. 133–140

14. I. Couso, S. Montes and P. Gil. The necessity of the strong alpha-cuts of a fuzzy set, Int. J. Unc., Fuzz. Knowledge-Based Syst. 9, 249–262 (2001)

15. I. Couso, S. Montes P. Gil. Second-order possibility measure induced by a fuzzy random variable, in C. Bertoluzza, M. A. Gil, and D. A. Ralescu (Eds.) *Statistical Modeling, Analysis and Management of Fuzzy data*, Springer, Heidelberg pp. 127–144 (2002)

16. I. Couso, L. Sánchez, Higher order models for fuzzy random variables. Fuzzy Sets and Syst. **159**, 237–258 (2008)

17. I. Couso, L. Sánchez, P. Gil, Imprecise distribution function associated to a random set. Inf. Sci. **159**, 109–123 (2004)

18. G. de Cooman, A behavioural model for vague probability assessments. Fuzzy Sets Syst. **154**, 305–358 (2005)

19. G. de Cooman, D. Aeyels, Supremum preserving upper probabilities. Inf. Sci. **118**, 173–212 (1999)

20. G. de Cooman, P. Walley, An imprecise hierarchical model for behaviour under uncertainty. Theor. Decis. **52**, 327–374 (2002)

21. T. Denœux, Modeling vague beliefs using fuzzy-valued belief structures. Fuzzy Sets Syst. **116**, 167–199 (2000)

22. T. Denœux, Maximum likelihood estimation from fuzzy data using the EM algorithm. Fuzzy Sets Syst. **183**, 72–91 (2011)

23. P. Diamond, Fuzzy least squares. Inf. Sci. **46**, 141–157 (1988)

24. P. Diamond, Interval-valued random functions and the kriging of intervals. Math. Geol. **20**, 145–165 (1988)

25. P. Diamond, Fuzzy kriging. Fuzzy Sets Syst. **33**, 315–332 (1989)

26. P. Diamond, P. Kloeden. *Metric Spaces of Fuzzy Sets,* World Scientic (Singapore, 1994)

27. D. Dubois, L. Foulloy, G. Mauris, H. Prade, Probability-possibility transformations, triangular fuzzy sets, and probabilistic inequalities. Reliable Comput. **10**, 273–297 (2004)

28. D. Dubois, J. Lang, H. Prade. Possibilistic logic. In D.M. Gabbay, C.J. Hogger, J.A. Robinson, D. Nute (eds.) *Handbook of Logic in Artificial Intelligence and Logic Programming*, Vol. 3, Oxford University Press, pp. 439–513 (1994)

29. D. Dubois, H. Prade, Evidence measures based on fuzzy information. Automatica **21**, 547–562 (1985)

30. D. Dubois, H. Prade, The mean value of a fuzzy number. Fuzzy Sets Syst. **24**, 279–300 (1987)

31. D. Dubois, H. Prade, *Possibility Theory* (PLenum Press, New-York, 1988)

32. D. Dubois, H. Prade, When upper probabilities are possibility measures. Fuzzy Sets Syst. **49**, 65–74 (1992)

33. D. Dubois, H. Prade, Fuzzy sets—a convenient fiction for modeling vagueness and possibility. IEEE Trans. Fuzzy Syst. **2**, 16–21 (1994)

34. D. Dubois, H. Prade, The three semantics of fuzzy sets. Fuzzy Sets Syst. **90**, 141–150 (1997)

35. D. Dubois, H. Prade, Gradual elements in a fuzzy set. Soft Comput. **12**, 165–175 (2008)

36. D. Dubois, H. Prade, Gradualness, uncertainty and bipolarity: making sense of fuzzy sets. Fuzzy Sets Syst. **192**, 3–24 (2012)

37. R. Féron. Ensembles aléatoires flous, C.R. Acad. Sci. Paris Ser. A 282, 903–906 (1976)

38. S. Ferson, L. Ginzburg, V. Kreinovich, H.T. Nguyen, S.A. Starks. Uncertainty in risk analysis: towards a general second-order approach combining interval, probabilistic, and fuzzy techniques, In *Proceedings of 2002 IEEE International Conference on Fuzzy Systems* (FUZZ-IEEE'02) 1342–1347. Honolulu, Hawaii, USA (2002)

39. S. Ferson, V. Kreinovich, L. Ginzburg, K. Sentz, D.S. Myers, *Constructing probability boxes and Dempster-Shafer structures, Sandia National Laboratories, SAND2002-4015* (Albuquerque, NM, USA, 2003)
40. Y. Feng, L. Hu, H. Shu, The variance and covariance of fuzzy random variables and their applications. Fuzzy Sets Syst. **120**, 487–497 (2001)
41. M.B. Ferraro, R. Coppi, G. González, A. Rodríguez, A. Colubi, A linear regression model for imprecise response. Int. J. Approximate Reasoning **51**, 759–770 (2010)
42. M.A. Gil, M.A. Lubiano, M. Montenegro, M.T. López, Least squares fitting of an affine function and strength of association for interval-valued data. Metrika **56**, 97–111 (2002)
43. G. González-Rodríguez, A. Blanco, A. Colubi, M.A. Lubiano, Estimation of a simple linear regression model for fuzzy random variables. Fuzzy Sets Syst. **160**, 357–370 (2009)
44. G. González-Rodríguez, A. Colubi, M.A. Gil, Fuzzy data treated as functional data. A one-way ANOVA test approach. Comput. Stat. Data Anal. **56**, 943–955 (2012)
45. J.A. Herencia, Graded sets and points: A stratified approach to fuzzy sets and points. Fuzzy Sets Syst. **77**, 191–202 (1996)
46. A. Herzig, J. Lang, P. Marquis. Action representation and partially observable planning using epistemic logic. In: Proceedings of the International Joint Conference on Artificial Intelligence (IJCAI-03), Acapulco, Morgan Kaufmann, San Francisco, pp. 1067–1072 (2003)
47. E. Hüllermeier. Learning from imprecise and fuzzy observations: data disambiguation through generalized loss minimization. Int. J. Approximate Reasoning (2013) http://dx.doi.org/10.1016/j.ijar.2013.09.003
48. M. Ishizuka, K.S. Fu, J.T.P. Yao, Inferences procedures and uncertainty for the problem-reduction method. Inf. Sci. **28**, 179–206 (1982)
49. E.P. Klement, M.L. Puri, D.A. Ralescu, Limit theorems for fuzzy random variables. Proc. Roy. Soc. London A **407**, 171–182 (1986)
50. R. Körner, On the variance of fuzzy random variables. Fuzzy Sets Syst. **92**, 83–93 (1997)
51. V. Krätschmer, A unified approach to fuzzy random variables. Fuzzy Sets Syst. **123**, 1–9 (2001)
52. R. Kruse, On the variance of random sets. J. Math. Anal. Appl. **122**, 469–473 (1987)
53. R. Kruse, K.D. Meyer, *Statistics with vague data* (D. Reidel Publishing Company, Dordrecht, 1987)
54. H. Kwakernaak, Fuzzy random variables definition and theorems. Inf. Sci. **15**, 1–29 (1978)
55. K. Loquin, D. Dubois, A fuzzy interval analysis approach to kriging with ill-known variogram and data. Soft Computing, Special issue on Knowledge extraction from low quality data: theoretical, methodological and practical issues **16**, 769–784 (2012)
56. M.A. Lubiano. Variation measures for imprecise random elements, Ph.D. Thesis, Universidad de Oviedo, Spain (1999). (In Spanish)
57. E. Miranda, G. de Cooman, I. Couso, Lower previsions induced by multi-valued mappings. J. Stat. Plan. Infer. **133**, 173–197 (2005)
58. E. Miranda, I. Couso and P. Gil. Study of the probabilistic information of a random set. In *Proceedings of Third International Symposium on Imprecise Probabilities and Their Applications* (ISIPTA'03). Lugano, Switzerland (2003)
59. E. Miranda, I. Couso, P. Gil, Random sets as imprecise random variables. J. Math. Anal. Appl. **307**, 32–47 (2005)
60. S. Moral. Comments on "Statistical reasoning with set-valued information: ontic versus epistemic view" by Inés Couso and Didier Dubois, Int. J. Approximate Reasoning (2014) http://dx.doi.org/10.1016/j.ijar.2014.04.004
61. H.T. Nguyen, On random sets and belief functions. J. Math. Anal. Appl. **65**, 531–542 (1978)
62. M. Puri, D. Ralescu, Fuzzy random variables. J. Math. Anal. Appl. **114**, 409–422 (1986)
63. E. Ruspini, New experimental results in fuzzy clustering. Inf. Sci. **6**, 273–284 (1973)
64. L. Sánchez, I. Couso, Advocating the use of imprecisely observed data in genetic fuzzy systems. IEEE Trans. Fuzzy Syst. **15**, 551–562 (2007)
65. L. Sánchez, I. Couso, J. Casillas, in *A Multiobjective Genetic Fuzzy System with Imprecise Probability Fitness for Vague Data, 2nd International Symposium on Evolving Fuzzy Systems 2006 (EFS06)* (Ambleside, Lake District, UK, 2006), pp. 131–136

66. L. Sánchez, I. Couso, J. Casillas, Advocating the use of imprecisely observed data in genetic fuzzy systems. IEEE Trans. Fuzzy Syst. **15**, 551–562 (2007)
67. L. Sánchez, J. Otero and J. R. Villar. Boosting of fuzzy models for high-dimensional imprecise datasets. In: Proceedings of 11th Information Processing and Management of Uncertainty in Knowledge-Based Systems Conference (IPMU). Paris, France (2006)
68. L. Sánchez, M.R. Suárez, I. Couso. A fuzzy definition of Mutual Information with application to the design of Genetic Fuzzy Classifiers. In: Proceedings of International Conference on Machine Intelligence (ACIDCA-ICMI05). Tozeur, Tunisia, pp. 602–609 (2005)
69. L. Sánchez, M.R. Suárez, J.R. Villar, I. Couso, Mutual information-based feature selection and partition design in fuzzy rule-based classifiers from vague data. Int. J. Approximate Reasoning **49**, 607–622 (2008)
70. G.L.S. Shackle, *Decision, Order, and Time*, 2nd edn. (Cambridge University Press, Cambridge, 1969)
71. P. Smets, The degree of belief in a fuzzy event. Inf. Sci. **25**, 1–19 (1981)
72. SMIRE Research Group at the University of Oviedo. A distance-based statistical analysis of fuzzy number-valued data. Int. J. Approx. Reason. (2013) http://dx.doi.org/10.1016/j.ijar.2013.09.020
73. P. Walley, *Statistical reasoning with imprecise probabilities* (Chapman and Hall, London, 1991)
74. P. Walley, Statistical inferences based on a second-order possibility distribution. Int. J. Gen. Syst. **26**, 337–384 (1997)
75. R.R. Yager, Generalized probabilities of fuzzy events from fuzzy belief structures. Inf. Sci. **28**, 45–62 (1982)
76. J. Yen. Generalizing the Dempster-Shafer theory to fuzzy sets, IEEE Trans. Syst. Man Cybern. **20**, 559–570 (1990)
77. J. Yen, Computing generalized belief functions for continuous fuzzy sets. Int. J. Approximate Reasoning **6**, 1–31 (1992)
78. L.A. Zadeh, Probability measures of fuzzy events. J. Math. Anal. Appl. **23**, 421–427 (1968)
79. L.A. Zadeh, The concept of a linguistic variable and its application to approximate reasoning, Information Sciences, Part 1: 8, 199–249; Part 2: 8, 301–357. Part **3**(9), 43–80 (1975)
80. L.A. Zadeh. Fuzzy sets and information granularity, Advances in Fuzzy Set Theory and Applications. In: Gupta M.M., Ragade R.K. and Yager R.R. (eds.), North-Holland, Amsterdam, 3–18 (1979).
81. L.A. Zadeh, Fuzzy probabilities. Inf. Proc. Manag. **20**, 363–372 (1984)
82. L.A. Zadeh, Toward a generalized theory of uncertainty (GTU)-an outline. Inf. Sci. **172**, 1–40 (2005)

Solutions to Exercises

Exercises of Chapter 2

1. (a) Conjunctive interpretation.
   (b) Disjunctive interpretation.
   (c) Conjunctive interpretation.
   (d) Disjunctive interpretation.

2. (a) The proportion of students that speak 3 or more languages is between 0.1
   (at least, student no. 1 satisfies this condition) and 0.9 (student no. 3 only
   speaks two different languages.) Those proportions coincide with the lower
   and upper probabilities of the set $[3, \infty)$. In fact:

   $$P_*([3, \infty)) = P(\{s_i \in \Omega : \Gamma(s_i) \subseteq [3, \infty), \Gamma(s_i) \neq \emptyset\}) = P(\{s_1\}) = 0.1,$$
   $$P^*([3, \infty)) = P(\{s_i \in \Omega : \Gamma(s_i) \cap [3, \infty) \neq \emptyset\}) = P(\Omega \setminus \{s_3\}) = 0.9.$$

   (b) The minimum value of the expectation is $\min E(\Gamma) = 0.1 \cdot 4 + 0.9 \cdot 2 = 2.2$.
   The maximum is $\max E(\Gamma) = 0.1 \cdot 2 + 0.1 \cdot 3 + 0.8 \cdot 6 = 5.3$. The expected
   number of languages spoken is a number bounded by those numbers.
   (c) Let us consider the random variables $X = \min \Gamma$ and $Y = \max \Gamma$. Their
   respective expectations 2.2 and 5.3 (they do coincide with the minimum and the
   maximum of the expectation of Γ). Their respective variances are, therefore:

   $$\text{Var}(X) = 0.9 \cdot (2 - 2.2)^2 + 0.1 \cdot (4 - 2.2)^2 = 0.36, \text{ and}$$
   $$\text{Var}(Y) = 0.1 \cdot (2 - 5.3)^2 + 0.1 \cdot (3 - 5.2)^2 + 0.8(6 - 5.2)^2 = 2.01.$$

   Their half sum is $\frac{\text{Var}(X) + \text{Var}(Y)}{2} = \frac{0.36 + 2.01}{2} = 1.185$ and it measures the
   dispersion of the images of Γ. It belongs to the parametrized family of scalar
   variances defined by Lubiano Reference 34 of Chap. 2.
   (d) All we know about the actual variance of the number of languages spoken is
   that it belongs to the set of variances of the measurable selections of Γ. The
   minimum of this set coincides with the variance of the selection Z_1 defined as
   follows:

© The Author(s) 2014
I. Couso et al., *Random Sets and Random Fuzzy Sets as Ill-Perceived Random Variables*,
SpringerBriefs in Computational Intelligence, DOI: 10.1007/978-3-319-08611-8

$$Z_1(s_1) = 4, Z_1(s_2) = 3, Z_1(s_3) = 2, Z_1(s_i) = 3, i = 4, \ldots, 10,$$

whose variance is $\text{Var}(Z_1) = 0.2$. It is strictly lower than the variances of X and Y.

On the other hand, the maximum of this set is the variance of one of the "most dispersed" measurable selection of Γ. There are several selections with the maximum variance. One of them is the random variable Z defined as follows:

$$Z_2(s_1) = 6, Z_2(s_2) = Z_2(s_3) = Z_2(s_4) = Z_2(s_5) = Z(6) = 2, Z_2(s_i) = 2,$$
$$i = 7, \ldots, 10,$$

whose variance is $\text{Var}(Z_2) = 4$. It is strictly greater than the variances of X and Y and, therefore, also greater than their half sum. According to this example, the minimum value of the variances of the selections of a random interval can be strictly lower than the variances of both extremes, as well as the maximum variance can be strictly higher than the maximum of the variances of both extremes. In general, the pair of variances of the extremes do not determine any of the bounds for the set of possible values of the actual variance.

3. Let us consider an arbitrary Borel measurable set $A \in \beta_{\mathbb{R}}$. Let us consider the dense subset of A, $D = \mathbb{Q} \cap A$. Let us consider the multi-valued mapping $\text{Int}(\Gamma) : \Omega \to \wp(\mathbb{R})$ that assigns, to each $\omega \in \Omega$ the interior of $\Gamma(\omega)$, i.e., the open interval $\text{Int}(\Gamma)(\omega) = (T_n(\omega), T_x(\omega))$. The upper inverse of A can be expressed as follows:

$$\Gamma^*(A) = \{\omega \in \Omega : \text{Int}(\Gamma)(\omega) \cap A \neq \emptyset\} \cup T_n^{-1}(A) \cup T_x^{-1}(A).$$

Furthermore, we can easily check that $\text{Int}(\Gamma)(\omega) \cap A \neq \emptyset$ if and only if $\text{Int}(\Gamma)(\omega) \cap D \neq \emptyset$, for every $\omega \in \Omega$. Therefore, the upper inverse of the open random interval can be expressed as follows:

$$\text{Int}(\Gamma)^*(A) = \text{Int}(\Gamma)^*(D) = \cup_{d \in D} \text{Int}(\Gamma)^*(\{d\})$$
$$= \cup_{d \in D} \left[T_n^{-1}(-\infty, d) \cap T_x^{-1}((d, \infty)) \right].$$

Therefore, $\Gamma^*(A) = (\cup_{d \in D}[T_n^{-1}(-\infty, d) \cap T_x^{-1}((d, \infty))]) \cup T_n^{-1}(A) \cup T_x^{-1}(A)$. Thus, we can express the upper probability of A in terms of the joint probability, as the probability that the 2-dimensional random vector takes values in the set:

$$(\cup_{d \in D}[(-\infty, d) \times (d, \infty)]) \cup (A \times \mathbb{R}) \cup (\mathbb{R} \times A).$$

4. Let us first assume that Γ is strongly measurable. The anti-image of any singleton $\{B\}$, for an arbitrary $B \in \wp(U)$, can be expressed in terms of the upper images as follows:
$$\Gamma^{-1}(\{B\}) = \cap_{C : C \subseteq B, C \neq \emptyset} \Gamma^*(C) \cap [\Gamma^*(B^c)]^c.$$

Conversely, let us now suppose that the set $\Gamma^{-1}(\{B\})$ is measurable for every $B \in \wp(U)$. Then, the upper inverse of any subset $A \in \wp(U)$ can be written as a finite union of sets of that form, $\Gamma^*(A) = \cup_{B \cap A \neq \emptyset} \Gamma^{-1}(\{B\})$, and therefore, it is measurable.

5. If X is measurable selection of Γ, then, $X(\omega) \in \Gamma(\omega), \forall \omega \in \Omega$ and therefore:

$$\Gamma(\omega) \subseteq A \Rightarrow X(\omega) \in A \Rightarrow \Gamma(\omega) \cap A \neq \emptyset, \ \forall A \in \wp(U).$$

Thus,

$$\Gamma_*(A) = \{\omega \in \Omega : \Gamma(\omega) \subseteq A\} \subseteq \{\omega \in \Omega : X(\omega) \in A\}$$
$$\subseteq \{\omega \in \Omega : \Gamma(\omega) \cap A \neq \emptyset\}, \ \forall A \in \wp(U),$$

and hence,

$$P_*(A) = P(\Gamma_*(A)) \leq P_X(A) \leq P(\Gamma^*(A)) = P^*(A), \ \forall A \in \wp(U).$$

In order to prove the second part, let us consider an arbitrary set $A \in \wp(U)$ and let us consider the following three disjoint families of sets:

$$\mathcal{F}_1 = \{B \in \wp(U) : B \cap A \neq \emptyset, B \cap A^c = \emptyset\} = \{B_1, \ldots, B_k\}$$
$$\mathcal{F}_2 = \{B \in \wp(U) : B \cap A \neq \emptyset, B \cap A^c \neq \emptyset\} = \{B_{k+1}, \ldots, B_l\}$$
$$\mathcal{F}_3 = \{B \in \wp(U) : B \cap A = \emptyset\} = \{B_{l+1}, \ldots, B_{2^{\#U}}\}.$$

Now, let us select an arbitrary element $b_i \in B_i$ for each $i \in \{1, \ldots, k\}$, two arbitrary elements $b_i^X \in B \cap A^c$ and $b_i^Y \in B \cap A$, for each $i \in \{k+1, \ldots, l\}$, and an arbitrary $b_i \in B$ for each $i \in \{l+1, \ldots, 2^{\#U}\}$.

Let us now define the mappings $X_A, Y_A : \Omega \to U$ as follows:

$$X_A(\omega) = \begin{cases} b_i & \text{if } \Gamma(\omega) = B_i, i \in \{1, \ldots, k\} \cup \{l+1, \ldots, 2^{\#U}\}, \\ b_i^X & \text{if } \Gamma(\omega) = B_i, i \in \{k+1, \ldots, l\}. \end{cases}$$

$$Y_A(\omega) = \begin{cases} b_i & \text{if } \Gamma(\omega) = B_i, i \in \{1, \ldots, k\} \cup \{l+1, \ldots, 2^{\#U}\}, \\ b_i^Y & \text{if } \Gamma(\omega) = B_i, i \in \{k+1, \ldots, l\}. \end{cases}$$

Since Γ is assumed to be strongly measurable, and according to the last exercise, we can easily check that the sets $X^{-1}(C)$ and $Y^{-1}(C)$ are measurable, for every $C \in \wp(U)$. Furthermore, $X_A^{-1}(A) = \Gamma_*(A)$ and $Y_A^{-1}(A) = \Gamma^*(A)$ and therefore, we have that $P_{X_A}(A) = P_*(A)$ and $P_{Y_A}(A) = P^*(A)$. Moreover, by construction, X_A and Y_A are selections of Γ.

(Let the reader notice that, when the images of Γ are singletons, we have that $\Gamma^{-1}(\mathcal{F}_2) = \emptyset$ and therefore X_A and Y_A do coincide. Otherwise, they do not, and, when, in addition, $P(\Gamma \in \mathcal{F}_2) \neq 0$, they induce different probability measures on U.)

6. (a) Let us consider the set function $m : \wp(U) \to [0, 1]$ defined as follows:

$$m(B) = P(\{\omega \in \Omega : \Gamma(\omega) = B\}) = P_\Gamma(\{B\}), \ \forall B \in \wp(U).$$

We observe that $m(\emptyset) = 0$ and $\sum_{B \in \wp(U)} m(B) = 1$, and therefore m is, formally speaking, a basic mass assignment. Furthermore, the upper and lower probabilities induced by Γ can be defined as functions of m as follows:

$$P^*(A) = P(\{\omega \in \Omega : \Gamma(\omega) \cap A \neq \emptyset\})$$

$$= \sum_{B : B \cap A \neq \emptyset} P(\{\omega \in \Omega : \mathcal{F}(\omega) = B\})$$

$$= \sum_{B : B \cap A \neq \emptyset} m(B),$$

$$P_*(A) = P(\{\omega \in \Omega : \Gamma(\omega) \cap A \neq \emptyset\})$$

$$= \sum_{B : B \subseteq A} P(\{\omega \in \Omega : \mathcal{F}(\omega) = B\})$$

$$= \sum_{B : B \subseteq A} m(B).$$

Therefore, P^* and P_* do respectively satisfy the properties of plausibility and belief functions.

(b) Given a belief function, $\mathrm{Bel} : \wp(U) \to [0, 1]$, there exists a unique mass assignment $m : \wp(U) \to [0, 1]$ such that $\mathrm{Bel}(A) = \sum_{B : B \subseteq A} m(B), \ \forall A \in \wp(U)$. Such a mass function is called the Möbius transform of Bel, and it can be expressed as follows [45] in terms on Bel. Let $F_1, \ldots, F_k \subseteq U$ denote the focal sets of m. Let us define the multi-valued mapping $\Gamma : [0, 1] \to \wp(U)$ as follows:

$$\Gamma(x) = \begin{cases} F_1 & \text{if } 0 \leq x < m(F_1) \\ \ldots \\ F_i & \text{if } \sum_{j=1}^{i-1} m(B_j) \leq x < \sum_{j=1}^{i} m(B_j) \\ \ldots \\ F_n & \text{if } \sum_{j=1}^{n-1} m(B_j) \leq x \leq 1. \end{cases}$$

We can easily to check that $P(\Gamma = F_i) = m(F_i), \ \forall i = 1, \ldots, n$.

7. The histogram associated to the original (precise) data:

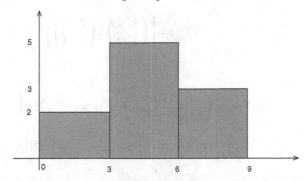

The "imprecise" histogram is the following one. It represents the collection of histograms where the respective heights are between the minimum and the maximum heights, and the sum of the three heights is equal to 10.

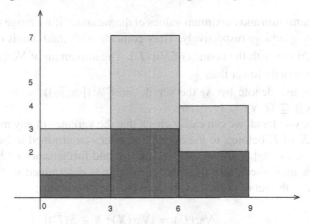

8. The maximum value for the variance (maximum of Kruse's variance) in both cases is reached, for instance, by the random variable X that respectively takes the values -10 and 10 for the outcomes h and t, respectively. This random variable is a measurable selection of both random sets. On the other hand, the minimum value is taken for any constant random variable, for instance, for the random variable $Y(h) = Y(t) = 0$, which is a measurable selection of both random sets. Moreover, for any value between 0 and 100 it can be defined a measurable selection of Γ_2 (and therefore, also a selection of Γ_1) with such a variance. Thus, Kruse's variance is equal to the interval $[0, 100]$ in both cases. On the other hand, the probability envelope of Γ_2 is included in the probability envelope associated to Γ_1, since $\Gamma_2(\omega) \subseteq \Gamma_1(\omega)$, $\omega = h, t$. Furthermore, the inclusion is strict. For instance, the probability measure that assigns probability 1 to the value 10 belongs to the envelope of Γ_1 but not to the probability envelope of Γ_2.

9. The probability envelopes of Γ and Γ' are, respectively,

$$\mathcal{P}(\Gamma) = \left\{ \left(\frac{1}{3}, \frac{2}{3} \right), \left(\frac{2}{3}, \frac{1}{3} \right) \right\}$$

and

$$\mathcal{P}(\Gamma') = \left\{ \left(\frac{i}{150}, \frac{150 - i}{150} \right) : i = 50, \ldots, 150 \right\}$$
$$= \left\{ (p, 1 - p) : \frac{1}{3} \leq p \leq \frac{2}{3}, \text{ and } 150p \in \mathbb{N} \right\}.$$

But both credal sets do coincide and they are the convex hull of both envelopes:

$$M(P_\Gamma^*) = M(P_{\Gamma'}^*) \left\{ (p, 1 - p) : \frac{1}{3} \leq p \leq \frac{2}{3} \right\}.$$

The minimum and maximum values of the variance, if we range the whole credal set are $\frac{2}{9}$ and $\frac{1}{4}$, respectively. They coincide with the bounds of $Var(\Gamma')$, but they do not with the bounds of $Var(\Gamma)$. The maximum of $Var(\Gamma)$ is equal to $\frac{2}{9}$ that is strictly lower than $\frac{1}{4}$.

10. Let us use denote by A_i the set $A_i = \Gamma^{-1}([a_i, b_i]) = \{\omega \in \Omega : \Gamma(\omega) = [a_i, b_i]\} \subseteq \Omega, \forall i = 1, \ldots, n$.

On the one hand, we can easily check that the variance of any measurable selection, X of Γ belongs to the interval of values determined in Eq. 2.19, because $X(\omega) \in [a_i, b_i], \forall \omega \in A_i, \forall i = 1, \ldots, n$ and furthermore $E(X) \in E(\Gamma)$.

Let us now check that the interval of values determined in Eq. 2.19 strictly includes the set of possible variances (Kruse's variance):

$$Var(\Gamma) = \{Var(X) : X \in S(\Gamma)\}.$$

We will prove that the value $\sum_{i=1}^n p_i \cdot \max\{(a_i - \max E(\Gamma))^2 p_i, (b_i - \min E(\Gamma))^2\}$ belongs to $\oplus_{i=1}^n [[a_i, b_i] \ominus E(\Gamma)]^2 \odot p_i$, but not to Kruse's variance. Let us consider the random variable $Z(\omega) = \max\{[a_i - \max E(\Gamma)]^2, [b_i - \min E(\Gamma)]^2, \forall \omega \in A_i, i = 1, \ldots, n$. We observe that $0 \leq (X(\omega) - E(X))^2 \leq Z(\omega), \forall \omega \in \Omega, \forall X \in S(\Gamma)$. Therefore $Var(X) \leq E(Z)$ and $Var(X) = E(Z)$ if and only if $[X(\omega) - E(X)]^2 = Z(\omega), c.s.(P)$. Let us suppose that there exists some measurable selection, $X \in S(\Gamma)$ such that $Var(X) = E(Z)$. On the other hand, since X is a selection of Γ, we have that $E(X) = \sum_{i=1}^n (p_{i1} a_i + p_{i2} b_i)$, where $p_{i1} + p_{i2} = p_i$. This expectation is strictly higher than $\min E(\mathcal{F})$ and strictly lower than $\max E(\Gamma)$, unless $p_{i1} = p_i, \forall i$ or $p_{i2} = p_i, \forall i$. According to this, $Var(X) < E(Z)$, unless one of those two alternatives happens. Let us suppose that $p_{i1} = p_i, \forall i$ (the proof under the second hypothesis is similar). In that case, $X = \max \mathcal{F}$ c.s.(P) and therefore there exists a collection of events

$C_1, \ldots, C_1$ such that $C_i \subseteq A_i$, $P(C_i) = P(A_i)$ and $X(\omega) = b_i$, $\forall \omega \in C_i$. Then we get a contradiction, if we suppose that $a_i < b_i$, for some $i \in \{1, \ldots, n\}$, because in that case, $\text{Var}(X) = \sum_{i=1}^{n} p_i (b_i - \max E(\Gamma))^2$ is strictly lower than $\max\{(a_i - \max E(\Gamma))^2, (b_i - \min E(\Gamma))^2\} = (a_i - \max E(\Gamma))^2$ and $E(X) = \max E(\Gamma)$.

11. (a) We have not been provided with the joint distribution of the random vector (X_0, Y_0), but we would expect them not to be independent (indeed, a positive correlation would be expected).

    (b) The random interval Γ_1 is a constant, and therefore it is stochastically independent from any other random set defined on Ω. In particular, it is stochastically independent from Γ_2.

12. Let us define the random sets Γ_1 and Γ_2 as follows:

$$\Gamma_1(i, j) = \begin{cases} \{\text{red}\} & i = 1, \ldots, 5, \\ \{\text{white}\} & i = 6, 7, \\ \{\text{red, white}\} & i = 8, 9, 10 \end{cases} \quad \forall j = 1, \ldots, 10.$$

$$\Gamma_2(i, j) = \begin{cases} \{\text{red}\} & j = 1, 2, 3, \\ \{\text{white}\} & j = 4, 5, 6, \\ \{\text{red, white}\} & j = 7, 8, 9, 10 \end{cases} \quad \forall i = 1, \ldots, 10.$$

We consider the Laplace probability measure on $\{1, \ldots, 10\} \times \{1, \ldots, 10\}$ that assigns probability 0.01 to each singleton $\{(i, j)\}$. The mass functions associated to the respective marginal probabilities are determined as follows:

$$P(\Gamma_1 = \{\text{red}\}) = 0.5, \ P(\Gamma_1 = \{\text{white}\}) = 0.2, \ P(\Gamma_1 = \{\text{red, white}\}) = 0.3$$
$$P(\Gamma_2 = \{\text{red}\}) = 0.3, \ P(\Gamma_2 = \{\text{white}\}) = 0.3, \ P(\Gamma_2 = \{\text{red, white}\}) = 0.4.$$

Furthermore, their joint probability is the product of both probability measures, and therefore they are stochastically independent.

The upper probability induced by the (Cartesian product) random set $\Gamma = \Gamma_1 \times \Gamma_2 : \{1, \ldots, 10\} \times \{1, \ldots, 10\} \to \wp(\{\text{red, white}\} \times \{\text{red, white}\})$ is determined by the upper probabilities of Cartesian products, that can be calculated as follows:

$$\begin{aligned} P_\Gamma^*(A \times B) &= P(\{(i, j) : \mathcal{F}(i, j) \cap A \times B \neq \emptyset\}) \\ &= P(\{(i, j) : \mathcal{F}_1(i, j) \cap A \neq \emptyset, \mathcal{F}_2(i, j) \cap B \neq \emptyset\}) \\ &= P(\{(i, j) : \mathcal{F}_1(i, j) \cap A \neq \emptyset\}) \cdot P(\{(i, j) : \mathcal{F}_2(i, j) \cap B \neq \emptyset\}) \\ &= P_{\Gamma_1}^*(A) \cdot P_{\Gamma_2}^*(B), \ \forall, A, B \subseteq \{\text{red, white}\}. \end{aligned}$$

According to the above information, the "marginal" upper probabilities are determined as follows:

A	{Red}	{White}	{Red,White}
$P^*_{\Gamma_1}(A)$	0.8	0.5	1

A	{Red}	{White}	{Red,White}
$P^*_{\Gamma_2}(A)$	0.7	0.7	1

The credal set associated to Γ is the set of probability measures dominated by P^*_{Γ},

$$\{P : P(C) \le P^*(C), \ \forall C \in \wp(\{red, white\} \times \{red, white\}\}.$$

13. The probability measure, $P_{(X_1,X_2)} : \wp(\{r, w\} \times \{r, w\}) \to [0, 1]$, associated to the joint experiment cannot be expressed as a product. In fact, there exists a stochastic dependence between the random variables X_1 and X_2, that represent the colours of both balls. Let us notice, for instance, that

- $P_{(X_1,X_2)}(\{(r, r)\}) = 0.15 + 0.2 \cdot \frac{1}{4} + 0.09 \cdot \frac{1}{6} + 0.12 \cdot \frac{1}{4}$
- $P_{X_1}(\{r\}) = 0.5 + 0.09 \cdot \frac{1}{6} + 0.09 \cdot \frac{5}{6} + 0.12 \cdot \frac{1}{2}$, and
- $P_{X_2}(\{r\}) = 0.3 + 0.2 \cdot \frac{1}{6} + 0.06 \cdot \frac{5}{6} + 0.12 \cdot \frac{1}{2}$

Thus, $P_{(X_1,X_2)}(\{(r, r)\}) = 0.245$ does not coincide with $P_{X_1}(\{r\}) \cdot P_{X_2}(\{r\}) = 0.65 \cdot 0.46$.

In the above expressions, r stands for "red" and w stands for "white".

14. The multi-valued mappings Γ_1 and Γ_2 defined as follows:

$$\Gamma_1(i) = \Gamma_2(i) = \begin{cases} \{r\} & i = 1, \dots, 5, \\ \{r, w\} & i = 6, \dots, 10. \end{cases}$$

represent our incomplete knowledge about the final colour of each ball. They are not independent. In fact:

$$P(\Gamma_1 = \{r\}) = P(\Gamma_2 = \{r\}) = P(\Gamma_1 = \{r\}, \Gamma_2 = \{r\}) = 0.5, \ \text{and}$$
$$P(\Gamma_1 = \{r, w\}) = P(\Gamma_2 = \{r, w\}) = P(\Gamma_1 = \{r, w\}, \Gamma_2 = \{r, w\}) = 0.5,$$

and therefore, $P(\Gamma_1 = A, \Gamma_2 = B) \ne P(\Gamma_1 = A) \times P(\Gamma_2 = B)$, in general.

15. The joint probability induced by the random vector (X_1, X_2) that represents the color of both balls can be factorized as the product of both marginals. In fact, the joint probability assigns the respective probabilities $\frac{9}{16}, \frac{3}{16}, \frac{3}{16}$ and $\frac{1}{16}$ to the respective possible outcomes (r, r), (r, w), (w, r) and (w, w).

16. In these conditions we have that:

$$P(\Gamma_i = [0, \infty)|\Gamma_{i+1} = [k, \infty)) = 1, \ \text{but} \ P(\Gamma_i = [0, \infty)|\Gamma_{i+1} \subseteq (-\infty, k)) = 0.$$

Therefore, Γ_{i+1} and Γ_i are not stochastically independent, $i = 1, \dots, n - 1$.

Exercises of Chapter 3

1. (a) Random fuzzy object. (Based on the ontic semantic of fuzzy sets.)

   (b) Ill-known random variable. (Based on the epistemic semantic of fuzzy sets.)

2. (a) The induced possibility distribution is defined as follows:

$$\pi(x) \begin{cases} 0 & \text{if } x \in (-\infty, -5] \cup [5, \infty) \\ 0.05 & \text{if } x \in [-5, -3) \cup (3, 5] \\ 0.35 & \text{if } x \in [-3, -2) \cup (2, 3] \\ 1 & \text{if } x \in [-2, 2]. \end{cases}$$

   (b) The degree of possibility of that interval is:

$$\Pi([d_0 - 5, d_0 - 2.5]) = \sup_{x \in [d_0 - 5, d_0 - 2.5]} \pi(x) = \pi(d_0 - 2.5) = 0.35.$$

3. It is easy to check, if we have into account that any trapezoidal fuzzy number is determined by four points in the plane, $(x_1, 0)$, $(x_2, 1)$, $(x_3, 1)$ and $(x_4, 0)$, and therefore it can be represented by the 4-dimensional vector (x_1, x_2, x_3, x_4). Thus, we can define a bijective function between the family of trapezoidal fuzzy sets and $\mathbb{R}^4$, $b : \mathcal{F}_T(\mathbb{R}) \to \mathbb{R}^4$. Let us consider the random vector $: \Omega \to \mathbb{R}^4$, defined as the composition of $\tilde{X}(\omega)$ with b. Since both random objects, $\tilde{X}$ and $\mathbf{X}$ are connected by means of a bijective function, the probability distribution of each of them is determined by the distribution of the other one.

4. The scalar variance can be expressed as a function of the variances of the mid points and spreads of the 10 outcomes of the fuzzy random variable. Taking into account that the quality of our information about each weight is the same for all the 10 objects, the variance of the spreads is null. Therefore, in this example, the scalar variance would just quantify the dispersion of the mid-points, i.e., the dispersion of the 10 displayed quantities.

   On the other hand, the fuzzy variance would be a fuzzy number representing the incomplete information about the variance of the actual weights of the 10 objects. Thus, the membership value $\text{Var}(\tilde{X})(x)$ would represent the degree of possibility that the actual variance coincides with x. We can equivalently say that, according to the available information, the α-cut $[\text{Var}(\tilde{X})]_\alpha$ contains the variance of the true weights with probability greater than or equal to $1 - \alpha$.

5. According to Exercise 3 of this chapter any triangular fuzzy random variable $\tilde{X}$ can be expressed as a bijective function of a 3-dimensional random vector $\mathbf{X} = b^{-1} \circ \tilde{X}$. Analogously, any pair of triangular fuzzy random variables $(\tilde{X}_1, \tilde{X}_2)$ can be expressed as a function of a 2-dimensional random vector $(\mathbf{X}_1, \mathbf{X}_2)$, whose components are, in turn, 3-dimensional random vectors. The joint probability of $(\tilde{X}_1, \tilde{X}_2)$ is univocally determined by the joint probability of $(\mathbf{X}_1, \mathbf{X}_2)$. Something similar can be said about the marginal distributions. Therefore, we can easily check that the joint probability of $(\tilde{X}_1, \tilde{X}_2)$ can be decomposed as the product of its marginals if and only if the joint probability induced by $(\mathbf{X}_1, \mathbf{X}_2)$ can be also decomposed as a product.

in the United States
asters